Elisabeth R. A. Cavalcanti Silva

Environmental assessment of areas vulnerable to desertification processes

AF319375

Elisabeth R. A. Cavalcanti Silva

Environmental assessment of areas vulnerable to desertification processes

A contribution to the theme of the transposition of the São Francisco River

ScienciaScripts

Imprint

Any brand names and product names mentioned in this book are subject to trademark, brand or patent protection and are trademarks or registered trademarks of their respective holders. The use of brand names, product names, common names, trade names, product descriptions etc. even without a particular marking in this work is in no way to be construed to mean that such names may be regarded as unrestricted in respect of trademark and brand protection legislation and could thus be used by anyone.

Cover image: www.ingimage.com

This book is a translation from the original published under ISBN 978-613-9-64352-3.

Publisher:
Sciencia Scripts
is a trademark of
Dodo Books Indian Ocean Ltd. and OmniScriptum S.R.L publishing group

120 High Road, East Finchley, London, N2 9ED, United Kingdom
Str. Armeneasca 28/1, office 1, Chisinau MD-2012, Republic of Moldova, Europe
Printed at: see last page
ISBN: 978-620-7-75811-1

DEDICATORY

I dedicate this work to my mother and grandmother for all the unconditional love they have shown me over the years.

ACKNOWLEDGEMENTS

I thank God, because without him I wouldn't be here and in him I place my faith, my trust and my love.

I would like to thank my advisor, Prof Dr Hernande Pereira da Silva, for his guidance, for which I am grateful. I mustn't forget to thank the teachers who, over the years, have passed on a fraction of their knowledge, motivating us and preparing us for the uncertainties of our chosen profession.

To all my friends for believing that dreams can come true if you just trust a little. Thanks especially to my friends Luis Eduardo, Cássia Luna, Cássia Herculano, Carol, Daiany and Monize, who have brightened my life so much during this period.

I am grateful for the availability of satellite images from INPE and the data from CONDEPE/FIDEM. I would also like to thank SERGEO for the use of ERDAS and ARCGIS software and GEOSERE for their support with the images.

Finally, I would like to thank IFPE for the opportunity it gave me, when it accepted me and helped me to expand my modest knowledge of the world, as well as the satisfaction I got from walking around its premises during the years that passed so quickly and that I miss. I couldn't forget the staff who, in one way or another, helped me reach the end of this first stage of my academic "career". So thank you all very much and may God bless us all. Amen.

I've built friends, faced defeat, overcome obstacles, knocked on life's door and told it: I'm not afraid to live it! Augusto Cury.

SUMMARY

Given the need for more academic studies on the transposition of the São Francisco River and its potential effects in terms of changes to the dynamics of the basins that will receive the water from this transposition, this study was carried out on the area corresponding to the middle section of the Ipojuca Basin, located in the state of Pernambuco, which will be directly influenced by possible environmental impacts from this project. This work uses image processing tools from the Landsat TM satellite to visualise the area in question, as well as to analyse the vegetation around the area. Based on flow data obtained from the Caruaru fluviometric station for the period 1973-2008, the aim was also to calculate the degree of vulnerability to climate change of the middle section of the Ipojuca Basin by combining relevant historical flow data, such as the non-stationarity of the historical flow series and the basin's climate variability. To this end, the Australian **Macro Water** *Sharing Plans* (MWSP) methodology will be used, as it also takes into account the human aspects of the area where it is applied, through the use ratio (RU) of water resources. By integrating the parameters of climate vulnerability and the ratio of use of water resources, the degree of hydrological stress in the Basin can be assessed in order to confirm and explain the existence of extensive areas of vulnerability and water scarcity, identified in satellite images, through vegetation and humidity indices. By integrating the results obtained using the MWSP methodology with the data obtained by calculating the NDWI moisture index for the same area and the SAVI vegetation index, it is clear that there are large areas of the middle section of the basin with a high degree of vulnerability and this study could help others by being able to combine hydrological models with data obtained through remote sensing in order to recognise the areas most in need of water. The results obtained show that there is high vulnerability of the basin to climate change, high hydrological stress in the basin and high levels of water stress in large areas in the middle section of the Ipojuca Basin. The results relating to high hydrological stress in the Basin show that there is a great need for water in the region surveyed and that the Basin has difficulties in meeting the need for water in the region. The results obtained for the high water stress in the Ipojuca Basin corroborate studies by the Ministry of National Integration of the São Francisco River Basin Integration Project with the Northern Northeast Basins, which already warned that various areas of the drier Northeast would face water stress that will worsen over the course of this century. In view of this, we conclude that there is a real need to transpose the waters of the São Francisco River to the study area, but we discuss the social benefits and potential environmental impacts that this transposition will bring to the area in question. Keywords: climate vulnerability, hydrological stress, water stress, environmental impacts.

SUMMARY

INTRODUCTION

The need to adopt effective water resource management tools to avoid or at least mitigate environmental impacts in river basins requires an understanding of the dynamics of natural processes related to the different factors that make up the different terrestrial landscapes. From this point of view, understanding patterns relating to climate, soil, vegetation, hydrography and land use are important components for drawing up a geoenvironmental analysis.

With the start of the industrialisation process, which began in England around the 18th century and later reached other regions of the planet, man began to take more and more of the natural resources available and, through industrial processes, transform them into manufactured products. This process has had many consequences, such as deforestation, water pollution from the dumping of industrial waste and air pollution from the emission of gases into the atmosphere (FREITAS & PORTO, 2006).

The 2007 IPCC (**Intergovernmental Panel on Climate Change**) report states that the rise in temperature observed since the middle of the 20th century is the result of increased concentrations of greenhouse gases in the atmosphere, caused by human activities. The report estimates a probability of more than 90 per cent that anthropogenic action on nature is the main cause of climate variations, which have become increasingly intense in recent years and are now known as climate change. For the IPCC, the term climate change refers to any change in the climate over a considerable period of time, regardless of whether it is a natural variation or the result of human activities.

According to the United Nations Educational, Scientific and Cultural Organisation UNESCO (2006), climate change is a long-term alteration caused by both natural factors and human activities. On the other hand, climate variability is a phenomenon inherent to the Earth's climate, in other words, it refers to the Earth's own dynamics (GALVÃO, 2008).

These changes in climate patterns can lead to phenomena such as the triggering or acceleration of desertification processes in more vulnerable areas. According to the United Nations Convention to Combat Desertification (UNCCD), desertification is understood as "land degradation in arid, semi-arid and dry sub-humid zones, resulting from various factors, including climatic variations and human activities", considering susceptible areas to be those with an aridity index between 0.05 and 0.65.

According to Souza Filho (2003), climate change has a direct influence on the water resources system and these changes in the climate will ultimately lead to the need to adopt management practices and instruments for these resources that adjust and adapt to the new climatic conditions.

These changes in climate influence not only the variation in hydrological behaviour but also the modification of natural elements that provide conditions for the sustainability of the natural environment, such as fauna and flora, giving rise to different environments based on the transformations that occur over time in the patterns of humidity, temperature and precipitation (TUCCI and MENDES, 2006; ALLAN, 1995).

With a view to measuring the susceptibility of river basins to a probable change in climate patterns and the need to know the various ways in which water resources are being used, various methods can be used for this purpose, including hydrological methods (GALVÃO, 2008).

These methods have the advantage of being simple, cheap, easy to apply and basically require hydrological data collected at river gauging stations, which are often the only ones available for the region under study. The methods in this category use tools from statistical hydrology, such as the mean, median and permanence curve to provide recommendations for the minimum guaranteed flow (GONÇALVES, 2003).

However, they are highly specific to the location and species for which they were developed, and therefore have limitations. They should therefore only be used as a preliminary survey for managing water resources and identifying critical points in river basins (ALVES & HENRIQUES, 1994).

In this sense, studies on the generation of climate forecast scenarios based on water balance modelling can help in the management of water resources in river basins, thus making it possible to propose alternatives a few years or even decades in advance, with a view to making better use of the available water resources (TUCCI, 1998).

It is important to note that the quantification of hydrological processes depends on the observation of variables that have random spatial and temporal behaviour and whose estimation depends on reliable and representative samples. A better understanding of the random behaviour of one or more variables that represent a water system ultimately depends on the information observed locally in this system (REIS **et. al.**, 2008).

According to Marengo (2006), due to the difficulty that models have in adequately representing the characteristics of the regional climate, the Brazilian experience shows the need to adjust the methods applicable to climate change scenarios resulting from global models for regional or local projections.

These methods can corroborate various Environmental Impact Assessment (EIA) studies in river basins that could potentially have their characteristics modified both by changes in the climate and by anthropogenic action, directly modifying the environment and causing environmental impacts.

According to the Ministry of National Integration (2011), scientific studies carried out in various areas of the Northeast have revealed that, in the short term, the drier Northeast will face water stress that will worsen throughout this century. For this and other reasons, the Ipojuca River Basin, which bathes several municipalities from the Sertão to the Pernambuco coast, will be one of the basins that will receive water from the transposition of the São Francisco River.

In this way, the analyses in this study are intended to provide information for an environmental assessment of part of the Ipojuca Basin (the middle section), with the aim of presenting new data that could help in understanding management, monitoring and conservation policies for the area in question, as well as discussing the real need to transpose the waters of the São Francisco River to this area.

OBJECTIVES

General Objective

To carry out an environmental assessment of the area of the middle section of the Ipojuca River Basin that will possibly have its dynamics affected by the Transposition of the waters of the São Francisco River.

Specific objectives

Av To statistically evaluate the degree of hydrological stress with the application of the Australian methodology called **Macro Water Sharing Plans** (MWSP) in the middle section of this Basin.

A To analyse vegetation water stress in the agreste part of the Ipojuca-PE basin using satellite images and vegetation indices.

=> Discuss the Ministry of National Integration's proposals for the Transposition of the São Francisco River.

=> To analyse the need for water in the study area and how the transposition of the waters of the São Francisco River could help the populations in the areas benefiting from the project, with an emphasis on the middle section of the Basin.

=> Identify the possible impacts of the São Francisco River transposition works on the study area.

CHAPTER 1

THEORETICAL BACKGROUND AND LITERATURE REVIEW

Several studies need to be and are being carried out in Brazil on the real need to transpose the waters of the São Francisco River and the environmental impacts resulting from it. Despite this, as far as academic research is concerned, there is still not much literature available on the subject, and most of the reliable data on the transposition works and the environmental impacts of the project in the areas that will receive water from the São Francisco River comes directly from the government agencies and sectors responsible for the project.

According to CONAMA Resolution 001 of 23/01/1986, environmental impact is defined as any alteration to the physical, chemical and biological properties of the environment caused by any form of matter or energy resulting from human activities that directly or indirectly affects: I - the health, safety and well-being of the population; II - social and economic activities; III - biota; IV - the aesthetic and sanitary conditions of the environment; and V - the quality of environmental resources.

An important stage in EIA is environmental diagnosis, which means describing and identifying environmental resources and their interactions, as they exist, in order to characterise the environmental situation of a given geographical area (current environmental quality), prior to the implementation of a given project (BRANDÃO and LIMA, 2002).

One of the aims of characterising the environmental situation (environmental diagnosis) is, like planning methodologies, to serve as a basis for understanding and examining the environmental situation, with a view to drawing up courses of action or making decisions to prevent, control and correct environmental problems (environmental policies and environmental management programmes) (SÁNCHEZ, 2006).

In this way, an environmental assessment presupposes the evaluation of a particular site with the application of different methodologies, with the aim of generating an environmental diagnosis, generally describing the levels of contamination, disturbance or preservation of a given area (EMBRAPA, 2010).

In this sense, an Environmental Impact Assessment (EIA) is essential in order to analyse the potential environmental impact that a large-scale project could have on a given river basin. According to Silva (1989), an Environmental Impact Assessment is a study process used to predict the environmental consequences resulting from the development of a project. This project could be, for example, the construction of a hydroelectric power station, large-scale irrigation, a harbour, a cement factory or a tourist resort, among others.

Thus, the licensing of activities that modify the environment, as defined in item I, article 48, of Decree-Law No. 32 of 18 November 1966; oil pipelines, gas pipelines, mining pipelines, sewage collection trunks and outfalls; electricity transmission lines above 230 KW; hydraulic works for the exploitation of water resources, such as: opening channels for navigation, drainage and irrigation, straightening watercourses, opening bars and inlets, transposing basins, dykes; fossil fuel extraction (oil, shale, coal); ore extraction, including class II, as defined in the Mining Code; landfills, processing and final disposal of toxic or hazardous waste; electricity generation plants, whatever the primary energy source, above 10 MW (SILVA, 1989).

The aim of the EIA is to ensure that the main problems can be foreseen at the planning stage of project development. The RIMA, in turn, presents the results of the technical and scientific environmental impact assessment studies, constituting a document of the environmental impact assessment process and clarifying all the elements of the proposal, so that they can be publicised by the social groups concerned and by all the institutions involved in decision-making. According to CONAMA resolution 001/86, the RIMA must clearly present the advantages and disadvantages of the project, as well as the environmental consequences of its implementation.

In this respect, Silva (1989) says that the RIMA must contain the project's objectives and justifications and their relationship and compatibility with sectoral policies; a description of the project and the technological and locational alternatives; a summary of the results of the environmental diagnostic studies; a description of the probable impacts of the implementation and operation of the activities; a characterisation of the future environmental quality of the area of influence of the project; a description of the expected effects of the mitigating measures in relation to the negative impacts; a programme for following up and monitoring the impacts and recommendations as to the most favourable alternative.

Environmental Impact Assessment (EIA) is based on Environmental Impact Studies (EIS). These studies are part of a set of technical and scientific activities that include environmental diagnosis, in order to identify, prevent, measure and interpret, where possible, environmental impacts.

In terms of Brazil, according to Verdum & Medeiros (1992), EIA came about as a result of demands from international funding bodies and was later incorporated as an instrument of national environmental policy in the early 80s. For Pimentel & Pires (1992), the EIA is not a decision-making instrument, but one that provides input for the decision-making process.

In the EIAs and RIMAs, which give rise to the EIA for the licensing required by law, three sectors are studied and focused on by multidisciplinary teams, with the aim of obtaining the scenario of that moment, so that a programme can be built that controls the multiple use of the natural resources involved. These are **Physical Environment** - studies climatology, air quality, noise, geology, geomorphology, water resources (hydrology, surface hydrology, physical oceanography, water quality, water use), and soil; **Biological Environment** - studies terrestrial, aquatic and transitional ecosystems; **Anthropic Environment - studies** population dynamics, land use and occupation, standard of living, production and service structure and social organisation. Therefore, EIA uses structured methods and techniques to collect, analyse, compare and organise data and information on environmental impacts in these three sectors (Costa et al. 2005).

According to Bastos and Almeida (2006), carrying out an environmental impact assessment generally follows these steps: developing a complete understanding of the proposed action; acquiring technical knowledge of the environment to be affected or studied; determining the possible impacts on environmental characteristics, quantifying the changes where possible; presenting the results of the analysis in such a way that the proposed action can be used in a decision-making process.

In order to present the main elements of the physical, biotic and socio-economic environment that could be modified by the implementation and operation of certain projects, it is beneficial to carry out an environmental diagnosis. Therefore, according to EMBRAPA (2010), making an environmental diagnosis means interpreting the problematic environmental situation of the area, based on the interaction and dynamics of its components, whether related to physical and biological elements or socio-cultural factors.

Therefore, as a way of trying to help in the environmental assessment of the areas that will receive water from the transposition of the São Francisco River, it is essential to try to obtain or generate more precise data on some of these areas, data that is often incipient or even non-existent.

And in order to demonstrate and quantify the needs of the study area, which in the case of this work refers to the middle section of the Ipojuca River Basin, and the possible environmental impacts of this project on the area, it is worth using new materials, as well as new techniques and technologies to better visualise the problems of the territory and the needs of the study area in view of the implementation of the project.

For Ross (2001), integrated studies of a given territory presuppose an understanding of the dynamics of how the natural environment functions with or without human intervention. In this sense, mapping the landscape units identified from the perspective of their fragility in the face of material conditions and possible human interventions is of valuable importance.

To this end, one of the techniques that can potentially be applied is geoprocessing. Novo (1989) cites some of the applications that this tool can offer in environmental assessment work, such as the assessment of water resources. Remote sensing data makes it possible to extend specific information to a wider spatial context and provides subsidies for the rational distribution of hydrological data collection points.

Rosa and Brito (1996) point out that the methods used to produce maps and geographical analyses are time-consuming and costly. It is therefore important to use new technologies such as information systems for basin planning. Satellite images provide a view of the complexities and dynamics of large areas of the earth's surface. You can then see the basin as a whole and its transformations, as well as the impacts caused by natural phenomena and human action in the use and occupation of space.

In this sense, the use of hydrological models to support observations of the dynamics of the basins studied is extremely important for quantifying problems related to possible changes in these dynamics.

Hydrological methods are characterised by establishing restriction flows using only data from historical flow series, on the understanding that this flow is sufficient to maintain certain characteristics of the ecosystem (GALVÃO, 2008). Restricted flow is, according to the National Water Agency (ANA, 2004), the flow that must be maintained in the river in order to ensure the maintenance and conservation of natural aquatic ecosystems, aspects of the landscape, as well as others of scientific and cultural interest.

In Australia, the New South Wales Department of Natural Resources has been using a methodology called **Macro Water Sharing Plans** (MWSP) since 2006 to determine the degree of hydrological stress in certain basins, taking as parameters to arrive at the final result the climatic vulnerability of the basin surveyed and the ratio of use of water resources in the area.

According to the method used by the MWSP, and in Brazil by the National Water Resources Plan (PNRH, 2006), and applied by Galvão (2008) in the Ribeirão Piripau Basin, it is possible to estimate hydrological stress through the ratio between extraction demand and available flow.
(called Hydrological Stress - Eh), which is the first indicator for estimating the Environmental Flow Potential, its parameters being, in addition to hydrological stress, Ecological and Cultural Value and Economic Dependence.

This methodology, which makes use of hydrological models, will be useful in the sense that it will serve as a parameter for the results obtained and observed in the calculation of the NDWI (normalised difference water index) from the LANDSAT-5 TM images from 1988, 1999, 2007 and 2011, on a scale of 1:100,000, analysing the water stress of the vegetation in the middle section of the Ipojuca basin, located in the agreste of the state of Pernambuco. This humidity index has already been used by authors such as Fensholt & Sandholt (2003) in their studies monitoring water stress in semi-arid environments in Senegal.

Freitas & Santos (2000) stated that surveys carried out by the United Nations World Organisation (WMO) indicate that one third of the world's population lives in regions of moderate to high water stress, and that statistics have clearly shown that over the next 30 years the global situation of water reserves will tend to worsen considerably if emergency measures are not taken to improve supply in relation to water demand.

According to Pereira et al. (1997), water stress is characterised when the soil does not contain water available to the plants, i.e. the evapotranspiration rate is more dependent on the soil's physical-water characteristics than on atmospheric demand and, in a second situation, when the soil contains available water but the plant is unable to absorb it at a sufficient rate and quantity to meet atmospheric demand (evaporative power of the air). This stress can be caused by factors such as: water deficit, low temperature, salinity, photoperiod or a combination of these factors.

As soon as water stress is triggered, changes begin to take place inside the plants, causing leaves to fall, reducing the leaf area, thus reducing the loss of water by the plants through evapotranspiration and increasing the water potential of the remaining leaves (CASTRO, 1994).

According to data from the Ministry of National Integration, the Environmental Impact Report (RIMA) for the Project to Integrate the São Francisco River Basin with the Basins of the Northern Northeast details the social, economic and environmental objectives of the project, which aims to ensure a water supply for the semi-arid northeast, whose population, without this alternative, will be forced to migrate. According to the Ministry of National Integration (2011), scientific studies show that, in the short term, the drier Northeast will face water stress that will worsen over the course of this century.

Currently, 1,340,000 km² of Brazilian territory are areas susceptible to desertification and cover the semi-arid tropics, dry sub-humid tropics and surrounding areas, directly affecting 30 million people. According to the Atlas of Areas Susceptible to Desertification in Brazil (MMA, 2007), of the total referred to above, 180,000 km2 are already in the process of severe and very severe desertification, "concentrated mainly in the Northeastern states, which have 55.25 per cent of their territory affected by varying degrees of environmental deterioration (apud CAMPOS, 2009).

Indicators of desertification include physical, biological and socio-economic processes such as erosion, salinisation and land use (MOUAT et al., 1997). According to CONTI (1998), desertification can be assessed by the aridity index and its indicators are: an increase in average temperature, a worsening of the soil water deficit, an increase in surface run-off, intensification of wind erosion, a reduction in rainfall, an increase in the daily temperature range and a decrease in the relative humidity (RH) of the air, caused by changes in climate patterns (apud NOLÊTO, 2005).

According to Galindo et al. (2008) the fundamental characteristic of the desertification phenomenon in the northeastern semi-arid region is the presence of patches of exposed soil. These are generally areas of shallow soil with no water retention capacity and physical and chemical limitations, which increase the ecological vocation for desertification. The soil is the conditioning factor in these more intensely degraded areas between higher and denser caatingas.

Changing the flow of part of the São Francisco riverbed to areas where the drought is more intense could represent greater economic and social dynamism for these areas, bringing development to entire regions. However, in the future it could lead to environmental impacts that will be difficult to overcome. For Sánchez (2006), the processes that generate environmental impacts derive from anthropogenic action or the combination of these at a given point.

Given the scarcity of material, it is necessary to obtain hydrological and spatial data and to discuss the benefits brought by the São Francisco River transposition project in the areas that will receive the water from this transposition, as well as to analyse the real need of these areas for water, with a view to economic and social use, hence the use of the MWSP methodology, which will be useful for this analysis.

Based on these studies, the parameters of vulnerability to climate change and the ratio of use of water resources will be analysed, which will be necessary to calculate the hydrological stress in the middle section of the Ipojuca basin located in the agreste region of the state of Pernambuco and, using the humidity index (NDWI), to assess the water stress of the vegetation in the basin.

CHAPTER 2

STUDY AREA

2.1 General characteristics of the Ipojuca River Basin

The Ipojuca Basin is located entirely in Pernambuco and lies between parallels 8° 09' 50" and 8° 40' 20" south latitude, and meridians 34° 57' 52" and 37° 02' 48" west longitude of the Greenwich meridian. Its area covers a surface of 3,433.58 km, corresponding to 3.49% of the state's total, and its perimeter is 749.6 km. The upper, middle and sub-middle sections of the basin are located in the Sertão (a small part) and Agreste regions of the state, while the lower section has most of its area situated in Pernambuco's Mata zone, including the state's coastal strip (CONDEPE, 2005) (Figure 1).

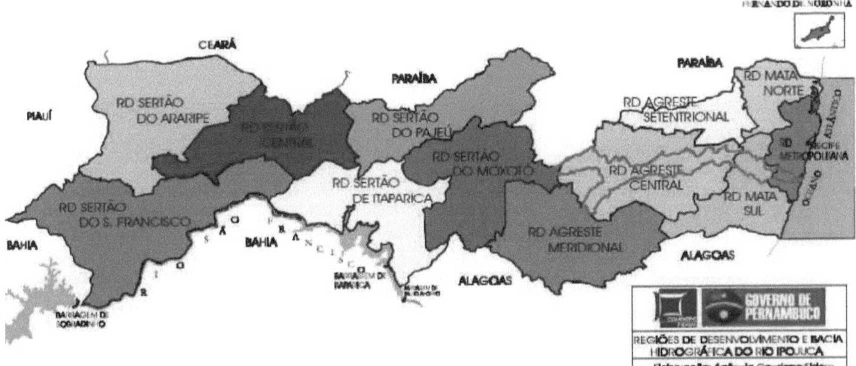

Figure 1 - Development Regions and the Ipojuca River Basin. Source: Basin Master Plan Hidrográfica do Rio Ipojuca - SECTMA. Prepared by the CONDEPE/FIDEM Agency.

Along this route, the area is recognised as a Water Planning Unit (UP 3) in the Pernambuco State Water Resources Plan (PERH/PE). The basin is bordered to the north by the Capibaribe river basin (UP2), to the south by the Una (UP 5) and Sirinhaém (UP 4) river basins; to the east with the second and third groups of small coastal river basins GL2 (UP 15) and GL3 (UP 16) and the Atlantic Ocean, and to the west with the Ipanema (UP 7) and Moxotó (UP 8) river basins and the state of Paraíba (CONDEPE, 2005) (ANNEX I).

The main river in the basin (Ipojuca River) rises on the slopes of the Pau d'Arco mountain range, in the municipality of Arcoverde, at an altitude of approximately 900m. Its course is orientated in a west-east direction, with an intermittent river regime until its middle course where it naturally becomes perennial in the municipalities of Gravatá and Chã Grande (CONDEPE, 2005) (Figure 2).

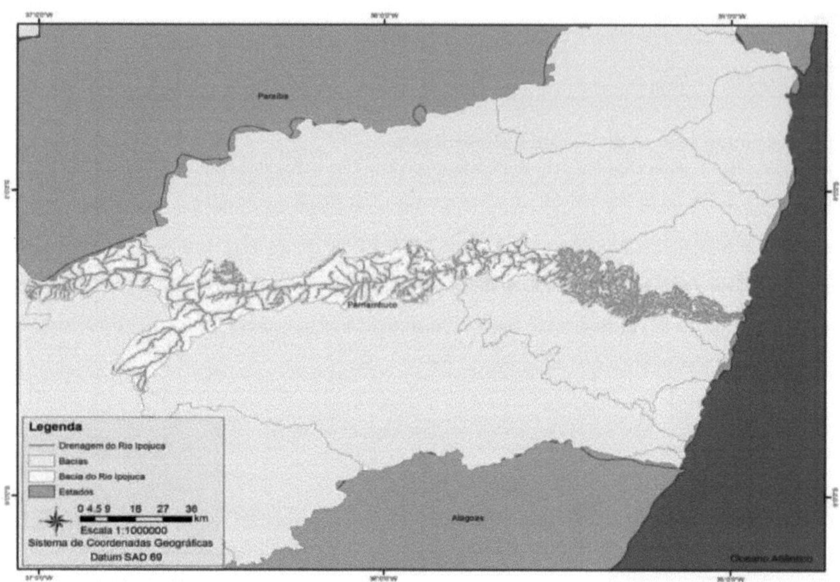

Figure 2 - Spatial location map of the Ipojuca-PE river basin.

The Ipojuca River bathes several municipal centres, including Sanharó, Belo Jardim, Tacaimbó, São Caetano, Caruaru, Bezerros, Gravatá, Primavera, Escada and Ipojuca (where its estuary has undergone enormous changes in recent years as a result of the installation of the Suape Industrial Port Complex). Pesqueira (606.79km^2), Caruaru (387.62km^2), São Caetano (262.37km2) and Sanharó (235.45km2) are the four municipalities with the largest areas belonging to the basin (CONDEPE, 2005) (ANNEX II).

2.1.1 Climate

The middle and sub-middle sections of the basin are located in the Agreste region of the state of Pernambuco. From a climatic point of view, this region is considered to be intermediate between areas with a humid climate (Zona da Mata) and a dry climate (Sertão), with characteristics of both. Thus, in the areas closest to the Sertão (upper and part of the middle stretches) the climate is hot and dry (semi-humid tropical climate), and the wettest period is from February to June (summer/autumn rains); in the sub-middle stretch (closer to the Zona da Mata), the rainy season extends from March to July (autumn/winter rains). The lower part of the basin is characterised by a hot and humid climate, with average rainfall of over 1,000mm per year, reaching over 2,000mm in the coastal areas. The rainy season lasts six months, from March to August (autumn/winter period) (CONDEPE, 2005).

2.1.1 Vegetation

The dominant vegetation shows physiognomic differences as a result of edaphoclimatic factors, and can generally be considered an "agrestina" caatinga, characterised by the presence of xerophytic, deciduous species, a large number of which are composed of thorns and an abundance of cacti and bromeliads (CONDEPE, 2005).

In the higher areas, exposed to the humid winds (the southeast trade winds), there are the "altitude marshes" (which stand out as spring areas) and are considered to be different ecosystems from those that predominate in the lower or less exposed areas. These marshes are home to highland forest, which is currently in a highly degraded state and is being replaced by polyculture. In the wetter areas of the basin, the vegetation is of the Atlantic Tropical Perennial Forest type, which has been greatly reduced by human activity. On the coast, mangroves are found, some of which are in the process of being devastated (CONDEPE, 2005).

2.1.2 Soils

In the upper, middle and sub-middle stretches of the Ipojuca river basin, according to CONDEPE (2005) and updated through the new soil classification, the soil classes Planossolos, Neossolos Regolíticos, Argissolos Amarelo e Vermelho-Amarelo and Neossolos Litólicos predominate and stand out, in addition to significant areas of Rock Outcrops (AR). Other soil classes are also found in these stretches, such as Fluvial Neosols and Latosols, but in less significant areas.

Table 1 shows that the land that can be used for annual and perennial crops, pastures, reforestation and wildlife, which correspond to classes II, III and IV, totals 1.280km^2 (37.30% of the basin's total area), while land unsuitable for intensive cultivation and adapted for pasture, reforestation and wildlife, but cultivable in cases of some special soil-protecting crops, which comprise classes V, VI and VII, reaches 1,868 km2 (54.43%). Land suitable only for the protection of wild flora and fauna, recreation or water storage (class VIII) totalled 284 km^2 (8.27%) (CONDEPE, 2005).

Table 1 - Summary of Land Use Capacity Classes in the Basin. Source: Ipojuca River Basin Master Plan - SECTMA. Prepared by the CONDEPE/FIDEM Agency.

Use Capacity Class(*) (**)	River basin sections				
	Top (km)^2	Average (km)^2	Sub-medium (km2)	Lower (km)^2	Total (km)^2
II	-	-	11	24	35
III	90	53	22	172	337
IV	448	308	134	18	908
V	299	128	139	129	695
VI	389	158	129	159	827
VII	149	30	75	92	346

| VIII | 120 | 75 | 71 | 18 | 284 |
| TOTAL | 1.495 | 752 | 573 | 612 | 3.432 |

(*)   The Ipojuca River Basin Master Plan adopted the classification scheme set out in the Manual for the Utility Survey of the Physical Environment and Land Classification in the Land Use Capacity System.
(**   ) Use capacity classes (I to VIII): based on the degree of use limitation;
• Class I: arable land with no conservation problems;
• Class II: arable land with simple conservation problems;
• Class III: arable land with complex conservation problems;
• Class IV: land that can be cultivated only occasionally or to a limited extent, with serious conservation problems.
• Class V: land generally suitable for grazing and/or reforestation, with no need for special conservation practices, cultivable only in very special cases;
• Class VI: land generally suitable for pasture and/or reforestation, with moderate conservation problems, cultivable only in special cases with some permanent soil-protecting crops;
• Class VII: land generally only suitable for grazing or reforestation, with complex conservation problems.
• Class VIII: land unsuitable for cultivation, grazing or reforestation, and can only be used for shelter and protection of wild fauna and flora, as a recreational environment, or for water storage purposes.

Table 2 shows Land Use and Occupation in the Ipojuca Basin by class. And

percentage of these classes in relation to the area in Km^2 .

Table 2. Land Use and Occupation in the Ipojuca River Basin. Source: Ipojuca River Basin Master Plan - SECTMA. Prepared by the CONDEPE/FIDEM Agency.

Class	Area (km)^2	%
Urban Area	28,89	0,84
Area under sugar cane cultivation	652,00	19,02
Forest area	193,78	5,64
Mangrove area	1,94	0,06
Area of open shrub and tree vegetation	14,00	0,41
Area of closed shrub and tree vegetation	351,84	10,24
Closed tree vegetation	246,20	7,17
Exposed soil	1,95	0,06
Anthropism	1.912,56	55,70
Weir	17,34	0,51
Unidentified use (cloud)	5,15	0,15
Unidentified use (shade)	6,93	0,20
Total	**3.433,58**	**100,00**

2.1.4 Water utilisation

Based on data relating to the characterisation of the Ipojuca river basin in relation to the different uses of water, carried out by the National Environment Programme - PNMA (2003) with water users, it was then possible, in terms of percentages, to see that there is a predominance of animal use (49.1%), followed by human consumption (20.8%), cleaning (18.1%), irrigation (6.6%), the industrial sector (4.6%) and, to a lesser extent, energy generation (0.7%), as can be seen in Figure 3:

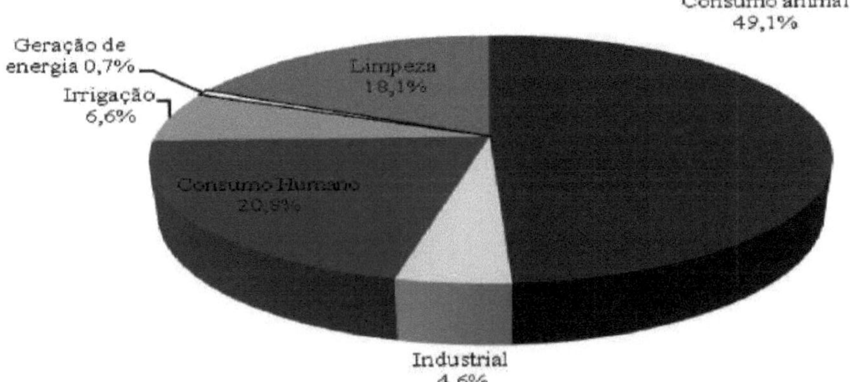

Figure 3 - Water use in the Ipojuca Basin. Source: PNMA (2003).

2.2 Transposition works on the São Francisco River

The Project for the Integration of the São Francisco River with the Hydrographic Basins of the Northern Northeast is a Federal Government undertaking, under the responsibility of the Ministry of National Integration. It is intended to ensure a supply of water, by 2025, to around 12 million inhabitants of 390 municipalities in the Agreste and Sertão of the states of Pernambuco, Ceará, Paraíba and Rio Grande do Norte (MINISTÉRIO DA INTEGRAÇÃO NACIONAL, 2011), (ANNEX III).

According to the Ministry of National Integration, the system will operate on both axes of the transposition. The North Axis, with 402 kilometres (km), will take water from Cabrobó (PE) to the Salgado and Jaguaribe rivers, in Ceará; Piranhas-Açu, in Paraíba and Rio Grande do Norte; and Apodi, also in Rio Grande do Norte. According to the project, the surplus volumes will be stored in strategic reservoirs in the receiving basins: Chapéu and Entre Montes (PE); Engenheiro Ávidos and São Gonçalo (PB); Atalho and Castanhão (CE); Armando Ribeiro Gonçalves, Santa Cruz and Pau dos Ferros (RN) (Figure 4) .

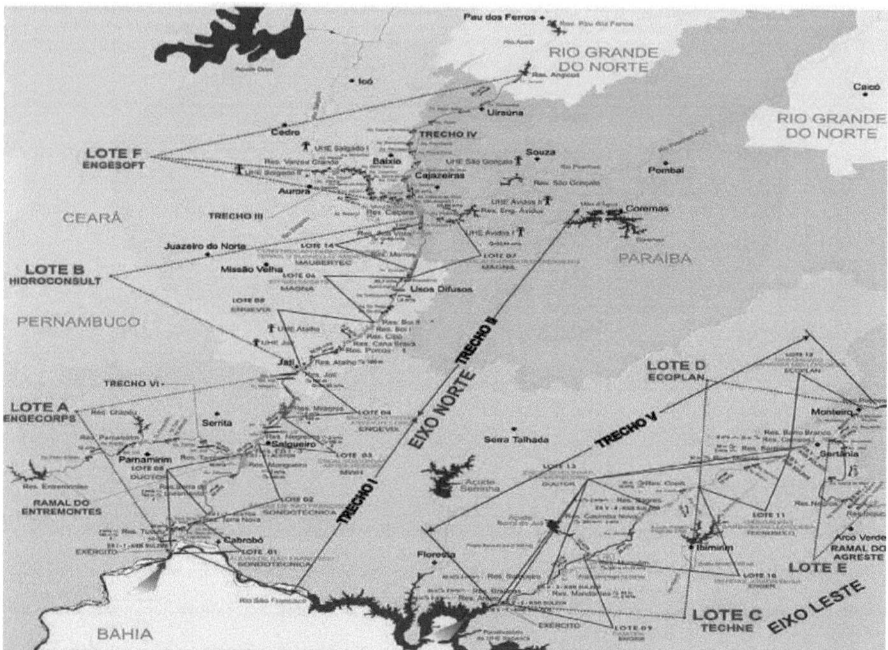

Figure 4 - Project for the Integration of the São Francisco River with the Hydrographic Basins of the Northern Northeast. Source: Ministry of National Integration, 2011.

In the Eastern Hub, the water will travel a distance of 220 kilometres. It will be captured in the lake of the Itaparica Dam (municipality of Floresta, Pernambuco), and will be carried to the Paraíba River (state of Paraíba), reaching the existing reservoirs in the receiving basins: Poço da Cruz, in Pernambuco, and Epitácio Pessoa (Boqueirão), in Paraíba. Part of the flow will be transferred beforehand to the Pajeú and Moxotó river basins and to the agreste region of Pernambuco, through the existence of a 70-kilometre branch that will connect the East Hub to the Ipojuca river basin. The maximum expected flow is 28 m³ /s, but the average operational flow will be 10 m³ /s. The surplus water will be transferred to the Poço da Cruz (PE) and Epitácio Pessoa (Boqueirão, PB) reservoirs (Figure 5).

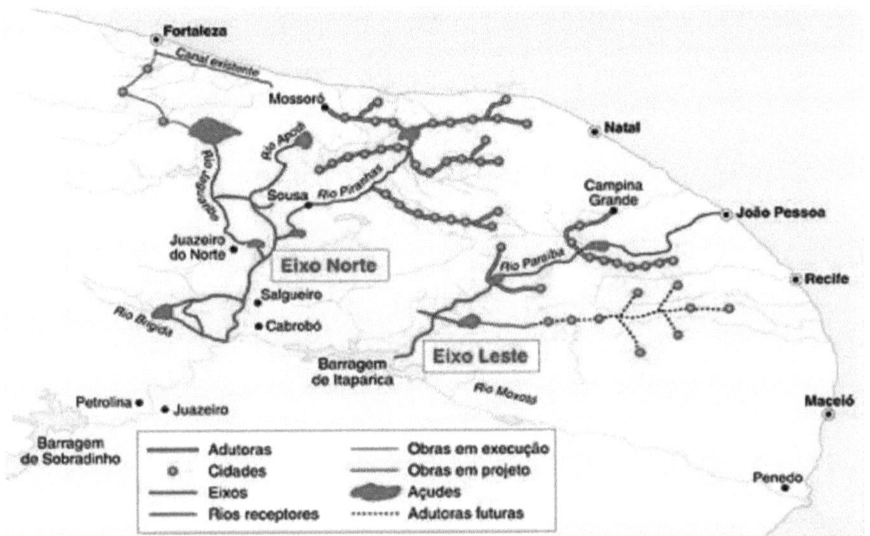

Figure 5 - Projected channels of the São Francisco River transposition. Source: Ministry of National Integration, 2011.

The engineering of the integration axes consists of open earth channels, with a trapezoidal section 25 metres wide and five metres deep, with a lining made up of an impermeable plastic membrane, covered with concrete. Aqueducts are being built on the stretches where rivers and streams are crossed, as well as tunnels to cross higher altitude areas.

To reach its destination, the water must overcome barriers imposed by the terrain. Nine pumping stations will be built to raise the water: three in the North Axis, to overcome altitudes of 165 metres, and six in the East Axis, where the water will be raised to an altitude of 304 metres. It is also planned to build 30 dams along the canals, which will act as compensation reservoirs to allow the water to flow even during the hours when the pumping is switched off (3 to 4 hours a day). Making the transposition and keeping the structure running are costly. This will make water more expensive for consumers. The final cost of water will be R$0.13 per 1000 litres ($m^3$). Source: (MINISTÉRIO DA INTEGRAÇÃO NACIONAL, 2011).

The integration of the São Francisco River with the temporary river basins of the semi-arid region will be possible with the continuous withdrawal of 26.4 m3/s of water, equivalent to just 1.42% of the flow guaranteed by the Sobradinho dam (1,850 m3/s), of which 16.4 m3/s (0.88%) will go to the North Axis and 10 m^3 /s (0.54%) to the East Axis (MINISTÉRIO DA INTEGRAÇÃO NACIONAL, 2011) (Figure 6).

19

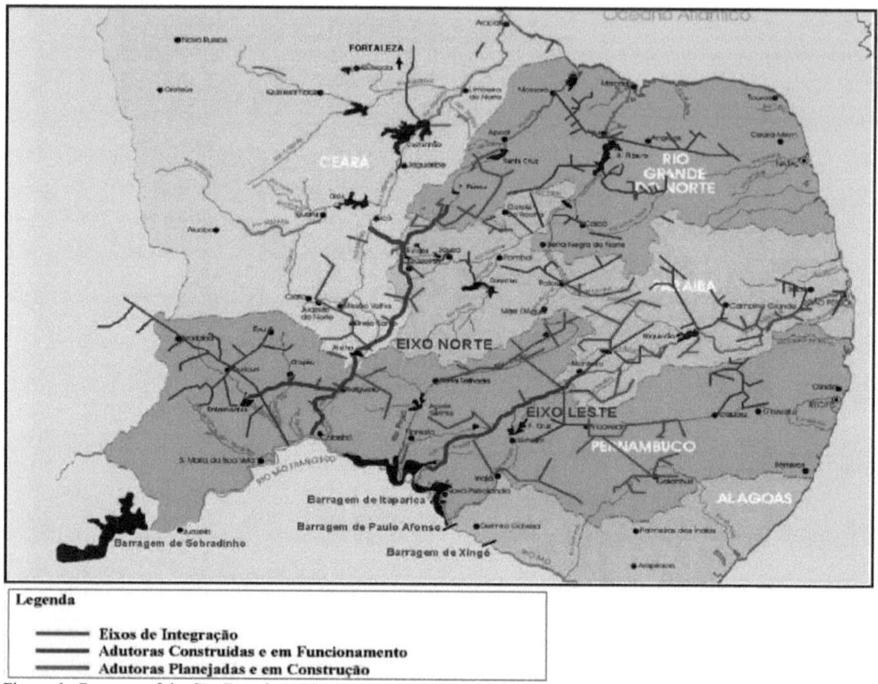

Figure 6 - Progress of the São Francisco River Transposition works. Source: Ministry of National Integration.

The 27 reservoirs that will receive the water transposed from the São Francisco River and the nine pumping stations will have an automatic and integrated monitoring system that will allow the volume of water at each point to be monitored and any reallocation from the fullest to the driest areas. An operation control centre will interconnect the water intake, the points of change in the direction of the canals, and the entrance to each state portal to allocate the surplus volume to strategic reservoirs (MINISTÉRIO DA INTEGRAÇÃO NACIONAL, 2011).

The Environmental Impact Report (RIMA) of the Project for the Integration of the São Francisco River Basin with the Basins of the Northern Northeast, published by the Ministry of National Integration, details the social, economic and environmental objectives of the São Francisco River Transposition project, which aims to ensure the supply of water to the semi-arid northeast.

The aim of the Transposition is to supply water for various purposes, but it should be noted that most of the water will be dedicated to irrigation (equivalent to 70 per cent of the water supplied), 26 per cent to industrial use and 4 per cent to the diffuse population. The transposition system is expected to be operational between 15 and 20 years from the start of construction.

The RIMA reported 44 expected environmental impacts of the project. Of these, 23 were considered the main ones, 11 positive and 12 negative (ANNEX IV). The main positive impacts include: an increase in available water and a reduction in losses due to reservoirs; the supply of up to 12.4 million people in cities, through urban supply systems that have already been set up, are being set up or are being planned by local authorities; the irrigation of abandoned areas and the creation of new agricultural frontiers, with the possibility, according to the studies carried out, of making approximately 161,500 hectares viable by 2025.500 hectares by 2025, 24,400 hectares of which will be for diffuse irrigation along the canals and 137,100 hectares for planned irrigation, and an increase in income and trade in the regions affected, since during the work there will be a big increase in trade and income in the towns that will house the construction sites, since in the long term the increase in employment and income will come from irrigated agriculture and industry, as a consequence of the transposition.

It is important to emphasise that the data on positive and negative impacts refers to the environmental impact report (RIMA) published by the government. According to the São Francisco River Basin Committee, this environmental impact report refers only to the axes to be implemented. Studies on the impacts on the donor basin, its tributaries and the receiving basins have not been carried out. And the absence of studies on the impacts at the mouth is also questioned.

One of the most worrying impacts is the process of desertification. Agenda 21 defined desertification as the degradation of land in arid, semi-arid and dry sub-humid regions, resulting from various factors, including climatic variations and human activities, where land degradation means the degradation of soils, water resources, vegetation and a reduction in the quality of life of the populations affected. Desertification is the degradation of dry land. This process consists of the loss of biological and economic productivity of agricultural land, pastures and native forest areas due to climate variability and human activities.

The desertification process manifests itself in two different ways: I) diffuse in the territory, encompassing different levels of degradation of soils, vegetation and water resources; II) concentrated in small portions of the territory, but with intense degradation of land resources.

In Brazil, the susceptible areas are located in the north-east and in the north of Minas Gerais. The susceptibility map of Brazil, drawn up by the MMA based on work carried out by IBAMA's Remote Sensing Centre, determined three categories of susceptibility: High, Very High and Moderate. The first two refer respectively to arid and semi-arid areas defined by the aridity index. The third is the result of the difference between the area of the Drought Polygon and the other categories. Thus, out of a total of 980,711.58 km^2 of susceptible areas, 238,644.47 km^2 are very highly susceptible, 384,029.71 km^2 are highly susceptible and 358,037.40 km2 are moderately susceptible (Figure 7).

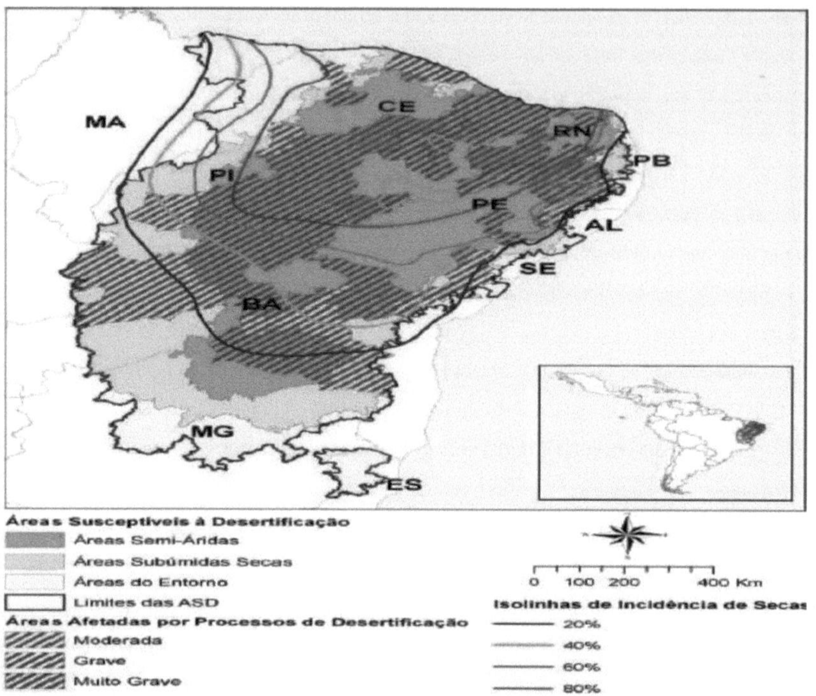

Figure 7 - Programme to combat desertification - PROÁGUA SEMIÁRIDO - Antidesertification. Source: OTAMAR, 2006.

The available studies indicate that the area affected in a Very Serious way is 98,595 km^2 , 10% of the semi-arid region, and the areas affected in a Serious way reach 81,870 km^2 , 8% of the territory. It should be added that the remaining areas subject to anthropism, 393,897 km2, suffer Moderate degradation.

In addition to these areas with diffuse levels of degradation, four areas with intense degradation can be mentioned, according to specialised literature, the so-called Desertification Cores. They are: Gilbués-PI, Irauçuba-CE, Seridó-RN and Cabrobó-PE, totalling an area of 18,743.5 km^2 .

Among the climatic regions covered by the UNCCD (arid, semi-arid and dry sub-humid), Brazil has no areas with an arid climate, only semi-arid and dry sub-humid (AI between 0.20 and 0.65). In addition to these classes, it was decided to add a third category to the SSA - the areas surrounding semi-arid and dry sub-humid areas. Although these areas do not formally fall within the climatic pattern considered susceptible to desertification, the reason for including them is that they have characteristics common to semi-arid and dry sub-humid areas. They also have a high occurrence of droughts and enclaves of vegetation typical of the Brazilian semi-arid region, the caatinga.

The criteria used to define the surrounding areas were as follows: a) surrounding municipalities that have been affected by droughts, in which case they are part of lists of municipalities officially assisted by drought emergency programmes administered by the Northeast Development Superintendence (Sudene); b) surrounding municipalities that are also part of the Caatinga Biome area, according to studies carried out by the National Council of the Caatinga Biome Biosphere Reserve (BEZERRA, 2004); and c) municipalities added to Sudene's area of operation, following the disciplining of Law no. 9.690 of 15 July 1998, such as those included in the state of Espírito Santo.

Vasconcelos Sobrinho, one of the pioneers in the study of desertification in the country, empirically selected six pilot areas where there were processes of soil and vegetation degradation in the states of Piauí, Ceará, Rio Grande do Norte, Paraíba, Pernambuco and Bahia. Based on its studies, the Ministry of the Environment organised field visits by a group of researchers to these areas. Among them, four were characterised as being at high risk of desertification, known as the Gilbués (PI), Irauçuba (CE), Seridó (PB) and Cabrobó (PE) Desertification Nuclei.

The RIMA also warns that some rivers will not be able to receive the projected volume of water, flooding parallel streams. Hence the concern about the forms of land use and occupation in the areas surrounding the Ipojuca Basin, because although the upper and middle stretches of the Basin have a great need for water, this Basin commonly has problems linked to pollution and flooding in its lower course. In the lower reaches, the need for water is not so great, as it is well known that the municipalities in this part of the Basin sometimes suffer from periodic flooding.

And it is in the lower reaches of the Basin that another major development is located, the Port of Suape, which also has an environmental impact on this area. In a study carried out by Silva **et. al.** (2009), land use and occupation were mapped using image fusion techniques from the HRC and CCD sensors of the CBERS 2B satellite, in an attempt to assess and delimit the spatial profiles of urbanisation in estuarine environments around the Port of Suape in 2007.

Figure 8 shows the environmental impact of the expansion of built spaces on natural environments, basically characterised by mangrove forests, through the growing number of industrial and port developments in the Suape area.

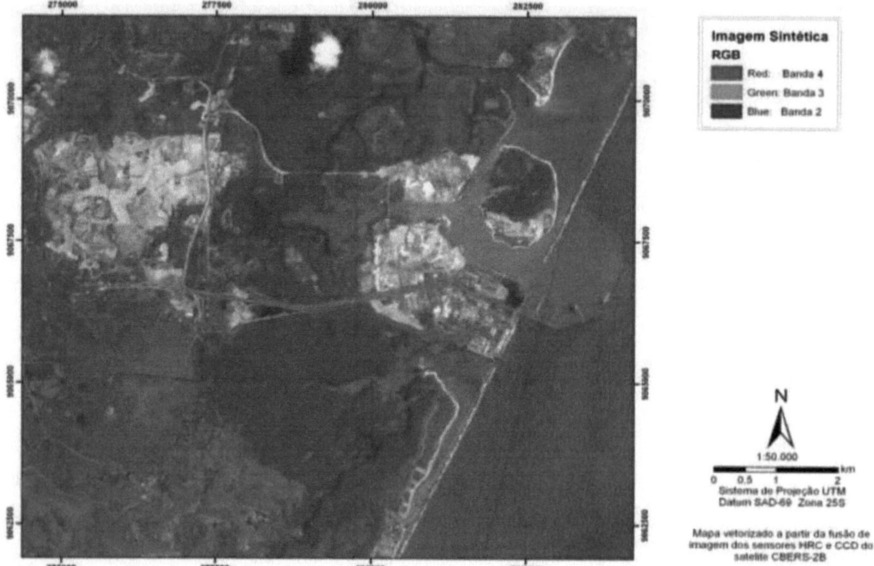

Figure 8 - Fusion of images from the HRC and CCD sensors of the CBERS 2B satellite, generating a synthetic image with a spatial resolution of 2.5m and in colour. Source: SILVA, et. al. (2009).

Using non-supervised analysis, seven main classes of land use and land cover were verified in the area around the port of Suape (as shown in Figure 9), with the most prominent being the area with exposed soil, a reflection of land occupation and the increase in the urban area, and extensive areas where polyculture and monoculture are practised, caused by the intense modifications brought about by the new developments. It was also possible to see extensive coverage of the surface by vegetation from the mangrove biome, although it is still possible to notice areas without vegetation in their central parts.

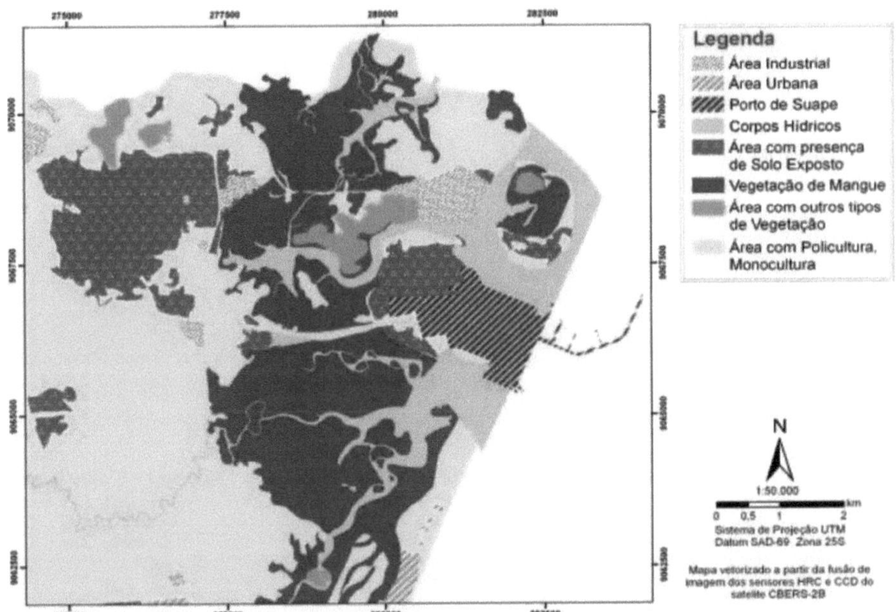

Figure 9 - Vectorisation of the product obtained using the fusion technique. Land Use and Occupation Map of the surroundings of the Port of Suape. Source: SILVA, et. al. (2009).

This shows the heterogeneity of this Basin, which is so important for the state of Pernambuco and especially for the semi-arid and agreste regions of the state, given the great need for water in these areas.

In this sense, we will try to evaluate the São Francisco River transposition project in the middle section of the Ipojuca Basin, which will have the natural dynamics of its basin affected by the works and the flow of water from the implementation of this project. This will generate new opportunities for land use and positive or negative impacts on nature in areas that did not receive water or received it at insufficient levels.

CHAPTER 3

MATERIAL AND METHODS

3.1 Data collected

This study used data obtained from the National Water Agency (ANA) through the fluviometric station called CARUARU, operated by the CPRM (coordinates: 8°12'39"S, 35°36'10" and Code 39340000).

It's important to note that some of the years in the fluviometric data used had to be discarded because they showed some inconsistencies: 1980, 1981, 1982, 1983, 1987, 1993 and 1994. There were also some years that did not have their flow data collected by CPRM, which were: 1995, 1996, 1997, 1998 and 1999.

The images used in this work are from the years 1988, 1999, 2007 and 2011, which were obtained from the TM (Thematic Mapper) sensor, point 214 and orbit 66 (21/06/1988), (28/02/1999), (19/07/2007) and (17/03/2011) on board the Landsat-5 satellite, obtained from the Image Generation Division of the National Institute for Space Research - INPE.

3.2 Calculating Hydrological Stress

In 2001, the Massachusetts Water Resources Commission carried out a study to define hydrological stress, which was defined as a basin or sub-basin in which the quantity of flow of a river has been significantly reduced or its quality degraded (MWRC, 2001).

The study by Galvão (2008) was used as a parameter for this work. It set out to develop a method for determining minimum environmental flows in unregulated rivers, taking into account the ecological, hydrological, social, economic and cultural aspects of the Ribeirão Pipiripau Basin (DF/GO).

The method developed in his work is a modification of the Australian **Macro Water Sharing Plans** Methodology - MWSP (NSW, 2006) of the Government of New South Wales - Australia. This method makes it possible to calculate the Environmental Flow Potential (EFP) for a management unit (river, stretch of river, etc.) and has three levels (low, medium and high corresponding to values 1, 2 and 3). PVA was defined as the product of three indicators: i) Hydrological Stress (Eh); ii) Ecological and Cultural Value (Vec); and iii) Economic Dependence (De).

The step taken to idealise this work was to calculate the hydrological stress in part of the Ipojuca Basin located in the agreste region of Pernambuco. Hydrological stress is the indicator that estimates the level of impact to which the watercourse is being subjected and is considered an important indicator of competition between water users.

It is calculated by balancing supply and demand, taking into account both the effects of water extraction and climate impacts on water bodies (GALVÃO, 2008). This indicator is the result of combining two parameters: Water Resources Use Ratio (Ru) and Vulnerability to Climate Change (Figure 10).

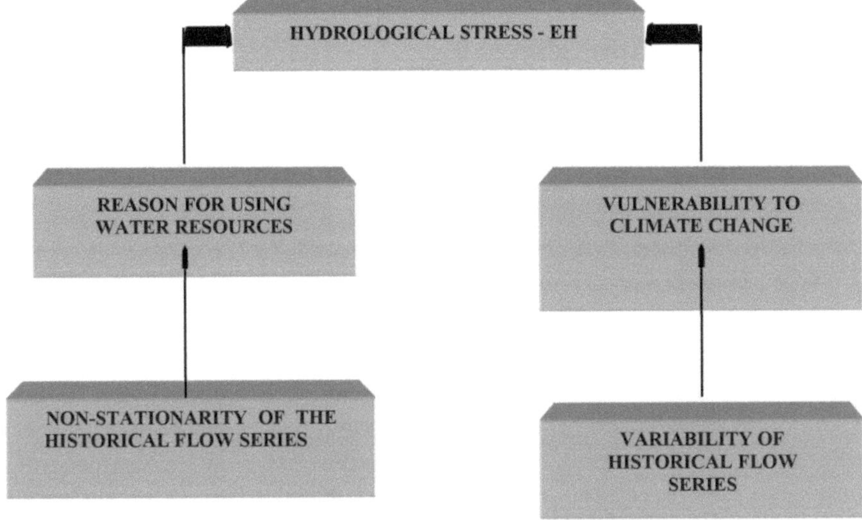

Figure 10. Parameters and sub-parameters that make up the Hydrological Stress - Eh indicator. Source: adapted from Galvão, 2008.

The calculation of hydrological stress proposed by Galvão (2008) was retained in this work and adapted to the area as a way of quantifying and subsidising the data obtained through remote sensing with the application of vegetation indices such as SAVI and the application of NDWI with a view to a better environmental assessment.

The results from the different methodologies applied will provide a more consistent environmental assessment to serve as a parameter for future environmental diagnosis.

3.2.1 Water Resources Use Ratio - Stage 1

The Water Resources Use Ratio parameter is an important measure of the impact of extraction, as it relates the amount of water available to the amount of water extracted from the watercourse, a similar criterion used by the MWSP and the PNRH (GALVÃO, 2008). This parameter takes into account the demand for extraction, which in this study will be considered as the demand for water for irrigation (m^3 /s) in the most critical month of the flow series studied, and the available flow as the long-term average of flows (Equation 1).

(Equation 1)

$$Ru = {Qd}/{Qmed} \; x \; 100$$

In which:

Ru = Water Resources Use Ratio

Qd = Demand (m3/s) in the most critical month

Qméd = Long-period average flow (m3/s)

Determining the average flow is important in a basin because it represents the maximum water availability. It is the highest flow that can be regularised, allowing the assessment of the upper limits of the use of water from a spring for different purposes (TUCCI, 2002).

Once the use ratio had been calculated, its value was examined: if the percentage of use of water resources is less than 20 per cent, its use ratio is considered low, if the value is in the range between 20 and 50 per cent, the use ratio is medium and if it is above 50 per cent, it is high, as shown below (Table 3).

Table 3 - Results for the Water Resources Use Ratio parameter in the Ipojuca Hydrographic Basin. Source: GALVÃO, 2008.

REASON FOR USE	% use of Ru
Bass	RU < 20%
Medium	**20 <RU<50%**
High	RU > 50%

The second parameter for measuring hydrological stress is the Vulnerability to Climate Change parameter, which characterises the vulnerability of the river basin to climate change, classifying it into three levels: low (B), medium (M) and high (A).

It is calculated using two sub-parameters that check the coefficient of variation of the historical flow series (variability) and the degree of stationarity of the historical flow series. These sub-parameters are climate variability and the non-stationarity of the historical flow series.

3.2.1 Climate Variability - Stage 2

The variability of hydrological conditions is a stochastic process in time and space due to the combination of various factors such as: climatic conditions of precipitation, evapotranspiration, solar radiation, among others; relief; geology, geomorphology and soils; vegetation cover and land use; anthropogenic actions on the river system (TUCCI and CLARKE, 2003 apud PAIVA, 2003).

Therefore, hydrological variability can be understood as the changes that may occur in the inputs and outputs of hydrological systems where the main inputs are: precipitation and evapotranspiration (which depends on other climatic variables), and the main output variable is: river flow (TUCCI, 2003).

Variability in this case is estimated by means of the coefficient of variation of the flows over the period studied. This is calculated using the standard deviation (SD), which is given by the formula below (Equation

2):

Equation 2

$$\sigma = \frac{R}{d2}$$

Where R is the average amplitude of the process and d2 is the factor related to the size of the subgroups. Once the historical series of average annual flows for the studied watercourse has been obtained, the coefficient of variation (CV) is calculated using Equation 3, where the standard deviation is divided by the average flow of the basin in the years studied and multiplied by 100.

Equation 3

$$cv = \frac{standard\ deviation}{average} . 100$$

The value found should be compared to those in Table 4, thus defining the level of variability for the management unit assessed. If the value found is less than 15 per cent, the value is low, if it is between 15 and 30 per cent, the value is medium and if it is greater than 30 per cent, it is high.

Table 4 - Variation of the sub-parameter Variability of the Historical Flow Series.

VARIABILITY	CV
Bass	< 15%
Medium	15% ≤ CV ≤30%
High	> 30%

High climate variability is associated with the occurrence of more severe extreme events than in regions with lower variability (SOUZA FILHO, 2003). As climate conditions change, the runoff from these basins also tends to change (TUCCI & MENDES, 2006). Therefore, high climatic and consequently hydrological variability makes the region more vulnerable to these events, necessitating strict water allocation and management criteria (GALVÃO, 2008).

The second sub-parameter to be assessed is the stationarity of the flow series, which can be estimated using the Salas test (1993). According to this test, a flow series is considered stationary if it is free of trends, variations or periodicities.

In other words, a stationary series means that the statistical parameters of the series, such as mean and variance, remain constant over time. Thus, it is assumed that the sequence of hydrological data, whether flow or precipitation, is statistically stationary in the sense that the values in the sequence fluctuate randomly around a mean value that remains constant over time, and that the dispersion of the data around the mean also remains constant. Otherwise, the series is considered non-stationary (CLARKE, 2003).

3.2.2 Non-stationarity of the historical flow series - Stage 3

The non-stationarity test for flow series is given by Equation 4 derived from Salas (1993):

Equation 4

$$tc = \frac{r\sqrt{N-2}}{\sqrt{1-r^2}} > t1 - \frac{\alpha}{2}, v$$

Where: tc (stationarity index) = value of t

r = correlation coefficient between q(i) (parameter) and number of years studied (i) (time)

N = number of years in the series

v = N- 2 degrees of freedom

t1-α/2,v = *Student*'s t at 99% confidence (tabulated value of Student's distribution for one degree of freedom n = N-2). The value obtained in the test is evaluated according to Table 5.

Table 5 - Variation of the Non-Stationarity sub-parameter of the Historical Flow Series.

NON-STATIONARITY	
LOW	tc < 0.9 t
MEDIUM	**0.9t≤tc≤ 1,1t**
HIGH	tc > 1.1 t

Where a tc less than 0.9t (0.9 x **Student's** t) represents low non-stationarity, i.e. either the series is stationary or its degree of non-stationarity is small. If the tc is in a range between 0.9t and 1.1t, it is considered to have medium non-stationarity and if it is above 1.1t (1.1 x **Student's** t), the historical flow series has a high degree of non-stationarity. It is important to emphasise that when tc is less than t, the series shows no significant upward or downward trend and is therefore considered stationary (CHAVES **et. al.**, 1997).

1.1.4 Climate Vulnerability - Stage 4

After finding the values of these two sub-parameters, they will be combined to create the parameter of vulnerability to climate change. These values (Table 6) were created using the three levels and their corresponding scores for each indicator (Low = 1, Medium = 2 and High = 3), parameter and sub-parameter. Each element in the table is the product of the scores corresponding to the levels in the row and column (GALVÃO, 2008).

Table 6 - Combination of the Variability and Non-Stationarity sub-parameters of the Historical Flow Series.

		NON-STATIONARITY	
		LOW	MEDIUM-HIGH
	LOW	1	23
VARIABILITY	MEDIUM	2	46
	HIGH	3	69

The integration of the previous Variability and Non-Stationarity sub-parameters (Table 6) enables the final calculation of the Vulnerability to Climate Change parameter (Table 7).

Table 7 - Variation in the Level and Score for the Vulnerability to Climate Change parameter.

VULNERABILITY TO CLIMATE CHANGE	SCORE
LOW	1 - 2
MEDIUM	3 - 4
HIGH	6 - 9

The higher the Variability and Non-Stationarity sub-parameters, the greater the vulnerability of ecosystems and the society that makes use of these resources to extreme events (GALVÃO, 2008).

By combining the parameters Water Resources Use Ratio and Vulnerability to Climate Change (Table 8), it is possible to obtain the Hydrological Stress indicator.

Table 8 - Combination of Water Resources Use Ratio and Vulnerability to Climate Change parameters.

		RATIO OF USE OF WATER RESOURCES - Ru		
		LOW	MEDIUM	HIGH
	LOW	1	2	3
Vulnerability to Climate Change	MEDIUM	2	4	6
	HIGH	3	6	9

3.2.5 Hydrological stress - Stage 5

Based on the work of Galvão (2008), the greater the Vulnerability to Climate Change and the Water Resources Use Ratio, the greater the Hydrological Stress - Eh and the more vulnerable the river basin is in terms of water supply and demand and susceptibility to severe climatic phenomena.

From the results obtained in Table 8, the relationship with the corresponding score is carried out, so that the final value for the Hydrological Stress - Eh indicator can be assessed (Table 9).

Table 9 - Variation in Level, Score and Final Value for the Hydrological Stress indicator - Eh.

LEVEL Eh	SCORE	VALUE
LOW	1 -2	1
MEDIUM	3-4	2
HIGH	6 - 9	3

Based on this methodology, and after working with the fluviometric data from the CARUARU station, located in the municipality of Caruaru, in the middle course of the Ipojuca basin corresponding to the agreste region of the state of Pernambuco, it was possible to calculate the hydrological stress that the watercourse in this study area has been experiencing.

3.3 Water Balance Model

Water balances were made for each of the stations in the Basin, using the Normal Water Balance by Thornthwaite & Mather (1955) in a spreadsheet developed by Rolim, G.S. & Sentelhas, P.C. (1998); Department of Physics and Meteorology, ESALQ-USP, 1998. This water balance was made for a series of 30 years.

3.4 UNEP aridity index

To identify areas susceptible to desertification, the UN proposed the Aridity Index - AI, which consists of the ratio between annual rainfall and potential evapotranspiration, the latter being calculated using the Thornthwaite method. The areas potentially susceptible to the process would be located within the AI of 0.05 to 0.065, which encompasses the arid, semi-arid and dry sub-humid climate zones (UNEP, 1991). The average annual potential evapotranspiration for each of the rainfall stations was calculated using the water balance.

Equation 5

$$Ia = \frac{P}{ETP}$$
(5)

P = average annual rainfall (mm)

ETP = Average annual potential evapotranspiration (mm)

According to the internationally accepted definition, the Aridity Index, defined as the ratio between Precipitation and Potential Evapotranspiration, establishes the following climatic classes (Table 10):

Table 10 - Aridity Index according to the Programme to Combat Desertification and Mitigate the Effects of Drought in South America.

Hyper-arid	**< 0,03**
Arid	0,03 - 0,20
Semi-arid	0,21 - 0,50
Dry sub-humid	0,51 - 0,65
Sub-humid humid	**> 0,65**

This index was used to establish risk areas and to draw up the World Atlas of Desertification, published by the UNEP and which serves as a parameter worldwide.

3.5 Software used

Excel software was used to draw up the figures for the fluviometric data and to tabulate the other research data. ERDAS 9.3 software was also used to process the images and ARCGIS 9.3 for spatial localisation and the assembly of maps and layouts of the Basin.

3.6 Image processing and layout assembly

Initially, all the images were registered from points collected in the field. To process the Landsat-5 satellite images, models were created using the **Model Maker** tool in the ERDAS Imagine 9.3 software and the final maps were assembled using the ArcGIS 9.3 software. Both are licensed by the Department of Geographical Sciences at the Federal University of Pernambuco.

3.6.1 Radiometric calibration

Radiometric calibration is obtained through the intensity of the radiant flux per unit of solid angle. Radiance represents the solar energy reflected by each pixel, per unit area, time, solid angle and wavelength, measured by the Landsat satellite in bands 1, 2, 3, 4, 5, 6 and 7 (OLIVEIRA, 2010). The set of radiance or radiometric calibration is obtained using the equation proposed by Markham and Baker (1987) (Equation 6):

Equation 6

$$L\,\lambda i = \alpha t + \frac{bt - \alpha t}{255}\ ND \tag{1}$$

Where a and b are the minimum and maximum spectral radiances, ND is the pixel intensity (integer between 0 and 255) and i corresponds to the bands (1, 2, 3, 4, 5 and 7) of the Landsat 5 and 7 satellite. The calibration coefficients used for TM images are those proposed by Chander and Markham (2003) apud Oliveira 2010.

3.6.2 - Reflectance

The reflectance (Equation 7) of each band (i) is defined as the ratio between the solar radiation flux reflected by the surface and the incident global solar radiation flux, obtained using the equation (Allen et al., 2002 apud Oliveira 2010):

Equation 7

$$\rho\,\lambda i = \frac{\pi\,.L\,\lambda i}{K\,\lambda i\,.\cos Z\,.dr} \tag{2}$$

Where λiL is the spectral radiance of each band, λik is the spectral solar irradiance of each band at the top of the atmosphere 1 2µm, Z is the solar zenith angle and rd is the square of the ratio between the mean Earth-Sun distance (ro) and the Earth-Sun distance (r) on a given day of the year (DSA) (Oliveira, 2010).

3.6.3   - *Normalised difference water index* (NDWI)

According to Cardozo et al. (2009), the NDWI vegetation index, proposed by Gao (1996), is related to the water content present in the leaves. Cardozo et al. (2009) state that this index is an important tool in studies related to vegetative vigour. It is obtained using the near infrared (pIR) and mid infrared (mIR) bands of the Landsat 5 satellite (OLIVEIRA, 2010) (Equation 8):

Equation 8

$$x = \frac{\rho\,NIR - \rho MidIR}{\rho NIR + \rho MidIR} \tag{3}$$

Studies show that NDWI is correlated with the water content of the plant canopy, indicating changes in biomass and oscillating moisture values in plants (HARDISKY et al., 1983; GAO, 1996).

3.6.4   - *Soil Adjusted Vegetation Index* (SAVI)

SAVI (Equation 9) was developed by HUETE (1988) as a transformation technique to minimise the influence of soil reflectance on spectral vegetation indices involving the red and near-infrared wavelengths and to more accurately model the near-infrared radiance in the more open canopies.

Equation 9

$$SAVI = \frac{(1+L)(\rho IV - \rho V)}{(L + \rho IV + \rho V} \tag{4}$$

CHAPTER 4

RESULTS AND DISCUSSION

4.1 Application of the MWSP methodology

Based on data from the Brazilian Institute of Geography and Statistics (IBGE) - Municipal Agricultural Production (2002) apud PERNAMBUCO (2004), the flow rate in m^3 per hectare for the municipalities with irrigated areas located in the Agreste region of Pernambuco that are part of the Ipojuca Basin was 19,000 m3/ha. Currently, the supply to Caruaru alone requires a continuous flow of close to 1,000 litres per second or 1m3/s, to serve a population of around 300,000 inhabitants.

This study analysed water consumption by the populations of the municipalities of São Caetano, Caruaru, Bezerros and Sairé, using water for irrigation. The information gathered on the consumption of irrigated areas serves as the basis for the first calculation to be carried out, which is the Water Resources Use Ratio, given by Equation 1. In it, the Ru value was approximately 31%. This leads to the conclusion that the Use Ratio was of an average standard, as shown in the table below (Table 11).

Table 11 - Hatched value of the variation in the Water Resources Use Ratio parameter.

REASON FOR USE	% use of Ru
Bass	RU < 20%
Medium	20 <RU<50%
High	RU > 50%

According to Ross (1992), climatic information, especially rainfall (intensity, volume, duration), is also important for analysing the potential and assessing the natural fragility of environments.

For this reason, a study by Salgueiro (2005) showed that precipitation in the Ipojuca Basin has a strong spatial dependence and that the Ipojuca Basin has low temporal variability of precipitation in relation to spatial variability, as demonstrated by analysing the years of medium, high and low variability.

Based on the data collected from the ANA at the CARUARU fluviometric station from 1973 to 2008, the average annual flow was calculated (Figure 11).

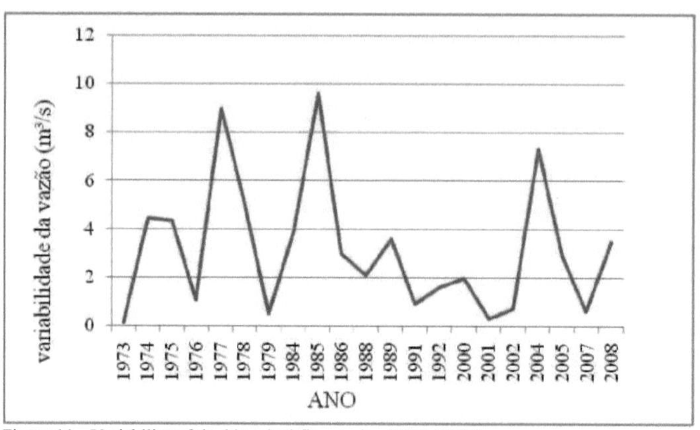

Figure 11 - Variability of the historical flow series (1973-2008).

Figure 8 shows that, according to the graph of flow variability in the years between 1973 and 2008, the flow in the Ipojuca Basin is quite variable hydrologically, but nothing can be said before the variability calculations that the MWSP methodology allows us to measure are carried out.

Next, the standard deviation (SD) and coefficient of variation (CV) were calculated, where the standard deviation (SD) is divided by the average flow of the years under study and multiplied by 100, resulting in a CV of around 86.59% (Table 12).

Table 12 - Average annual flow values in m³ /s provided at the station over the period (1973-2008).

YEAR	FLOW
1973	0,138183
1974	4,452208
1975	4,340556
1976	1,091182
1977	8,99389
1978	5,035929
1979	0,500255
1984	3,953351
1985	9,622021
1986	3,008478
1988	2,098384
1989	3,592545
1991	0,93164
1992	1,588845
2000	1,964701
2001	0,313202
2002	0,752737
2004	7,321267
2005	2,940758
2007	0,617488
2008	3,493978
AVERAGE	3,17865
D.P.	2,75247
C.V.	86,59%

Variability is therefore considered high in the middle section of the Ipojuca Basin because the CV is above 30% (Table 13).

Table 13 - Result (hatched) for the sub-parameter Variability of the Historical Flow Series in the Ipojuca River Basin.

VARIABILITY	CV
Bass	< 15%
Medium	**15% < CV <30%**
High	> 30%

Subsequently, the non-stationarity test (to find out if the series is stationary or not) was calculated using the formula in equation 4. It should be noted that the data obtained in this work has a sampling period of 21 years, and as the above test is based on **Student's** t-test at 99% confidence (tabulated value of Student's distribution for one degree of freedom n = N-2), the degree of freedom obtained was 0.0127. The r in the formula, which is the tendency to increase or decrease, is calculated using the flow data, resulting in a value of 0.033. The realised tc is therefore shown in the table (Table 14):

Table 14 - Calculated values for the SALAS method (1993).

AVERAGE	3,17865
D.P.	2,75247
C.V.	86,59%
T	0,0127
R	0,033
Tc	0,143922052

Thus, according to the values of tc in relation to 0.9t and 1.1t, tc has a value greater than 0.9t and 1.1t, and it can be concluded that the historical series shows a high degree of non-stationarity in the middle section of the Basin (Table 15).

Table 15 - Degree of non-stationarity.

tc	T(Student)	0,9t	1,1t	Non-stationarity
0,1439	0,0127	0,0114	0,0139	HIGH

Based on the above calculation (Table 14), the corresponding value should be hatched in the non-stationarity table below (Table 15). In the case of the middle section of the Ipojuca Basin, as discussed in the previous table, non-stationarity was considered high in the periods from 1973 to 2008, so the value of tc (which had a value above 1.1t) is considered high (Table 16).

Table 16 - Result (hatched) for the Non-Stationarity sub-parameter of the Historical Qmed Flow Series (1973-2008) in the Ipojuca River Basin.

NON-STATIONARITY	
LOW	tc < 0.9 t

| MEDIUM | $0.9t < tc < 1,1t$ |
| HIGH | $tc > 1.1\,t$ |

Non-stationarity can occur due to one or more of the following factors: climate variability during the sampling period, climate change or changes in land use (TUCCI & BRAGA, 2003).

In the method used, the values for the tables combining parameters or sub-parameters, as shown in Table 6, were created using the three levels and their corresponding scores for each indicator (Low = 1, Medium = 2 and High = 3), parameter and sub-parameter.

From these two sub-parameters (variability and non-stationarity) we can cross-reference (or integrate) them and calculate the Basin's Vulnerability to Climate Change parameter (Table 17).

Table 17 - Combination of the Variability and Non-Stationarity sub-parameters of the Historical Flow Series.

		NON-STATIONARITY	
		LOW	MEDIUM-HIGH
	LOW	1	23
VARIABILITY	MEDIUM	2	46
	HIGH	3	69

Both climate variability and non-stationarity were considered high, represented by the product of the two (3 x 3 = 9) as shown in Table 18:

Table 18 - Results (hatched) for the vulnerability to climate change parameter for the period 19732008.

VULNERABILITY TO CLIMATE CHANGE	SCORE
LOW	1 - 2
MEDIUM	3 - 4
HIGH	6 - 9

By integrating the parameters Use Ratio and Vulnerability to climate change (Table 19), the Basin's hydrological stress (Eh) can be assessed.

Table 19 - Combination of Water Resources Use Ratio and Vulnerability to Climate Change parameters.

		REASON FOR USING RESOURCES WATER - Ru		
		LOW	MEDIUM	HIGH
	LOW	1	2	3
Vulnerability to Climate Change	MEDIUM	2	4	6
	HIGH	3	6	9

As Ru was medium (2) and vulnerability was high (3), the product of these two is the number 6, so this number is compared with the level of Hydrological Stress, as shown in the table below (Table 20):

Table 20 - Result (hatched) of the Level Variation, Score and Final Value for the Hydrological Stress - Eh indicator.

LEVEL Eh	SCORE	VALUE
LOW	1 -2	1
MEDIUM	3 -4	2
HIGH	6 - 9	3

Therefore, by combining variability with non-stationarity, the following scores were obtained high (9) for vulnerability to climate change, and by combining the Use Ratio of water resources with vulnerability, a high overall score (6) was obtained for the hydrological stress indicator (Eh = 3).

It's worth pointing out that since the ratio of use of water resources was only analysed in a few municipalities in the middle section of the Basin, due to the lack of data on water consumption in the municipalities within its drainage area, the impact on the Ipojuca River Basin could be much greater than the results of this study, since Ru was considered average only due to the demand of a few municipalities, but even so, resulting in high hydrological stress in the Basin.

If we compare these results with those obtained by Galvão (2008) in the Piripau River Basin (DF/GO), we can see that this study used practically the same methodology as the author. The main difference between this work and his, in terms of methodology, was that he worked with data from various stretches of the basin.

A study carried out directly in the Basin in question by Silva et. al. (2009) analysed the Basin's situation in the face of a temperature increase of around 1°C. In the study, the authors carried out this simulation in order to assess how a slight increase in this parameter could affect surface runoff in the region under study. It is important to note that the simulated temperature value was even below the scenarios predicted by the IPCC (2001), which for the north-eastern region of Brazil indicates increases in temperature that could range from 1.4 to 5.8°C.

Through this study, Silva et. al. (2009) concluded that no trends in rainfall were observed, since the behaviour of evapotranspiration was as expected, i.e. it followed the trend of rising temperatures both during periods of lower water availability and during periods of higher water availability.

However, with regard to the flow results, it was observed that during the periods of highest flow in a temperature increase scenario, there is a tendency for it to decrease at all the stations studied (in total, five rainfall stations were used: Pesqueira, Caruaru, Poção, Sanharó and São Caetano. Each with a concomitant series of data between (19631992), with the exception of 1964, 1965, 1968, 1979 and 1984, totalling a series of 25 years of data), (SILVA et al., 2009).

At these stations, the flow for the scenario was always lower than normal, indicating that with an increase of 1° C, the flow would be lower. This result indicates a tendency for rivers to dry up earlier than normal. As for the periods of lower flow, there were variations, with periods simulating an increase in flow and others a decrease in flow (SILVA et. al., 2009).

Given the results obtained by Silva et. al.(2009) and in comparison with the results obtained in this study of high hydrological stress, it can be seen that, although the methodology adopted differs from that adopted by the aforementioned author and because it does not work with future trends, there were results that corroborate each other, since they indicate the vulnerability to climate change of the Basin in question. In this sense, there is some concern as this is a Basin that will receive water from the transposition of the São Francisco River, thus incorporating, in addition to the Basin's climatic susceptibility, negative impacts resulting from this transposition.

4.2 Application of Moisture Indices (NDWI) and Vegetation Indices (SAVI)

The first image to be analysed in this work is from 21/06/1988, in an area corresponding to the municipalities of São Caetano, Caruaru, Bezerros and Sairé, in the middle section of the Ipojuca Basin (Figure 12).

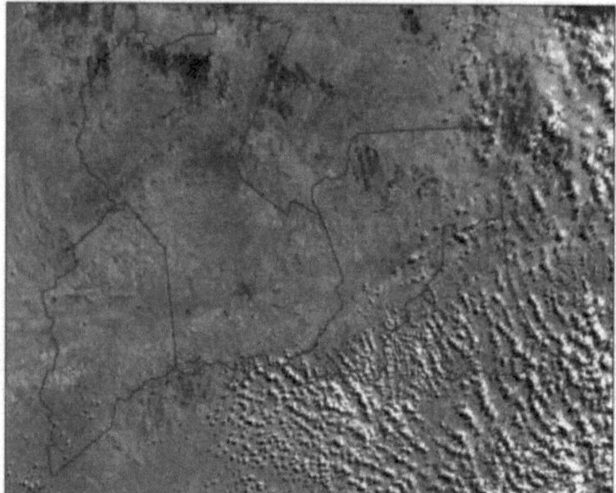

Figure 12 - Location of the municipalities in the middle section in 1988.

The **Normalised Difference Water Index** (NDWI), proposed by Gao (1996), is related to the water content present in the leaves. The advantage of using NDWI over other indices is that it is less sensitive to atmospheric effects. By applying the humidity index (NDWI) to the image above and dividing it into three classes according to the water content of the vegetation, it was possible to arrive at an NDWI result that indicates that the area covered by the research had vegetation with low to moderate water content (Figure 13).

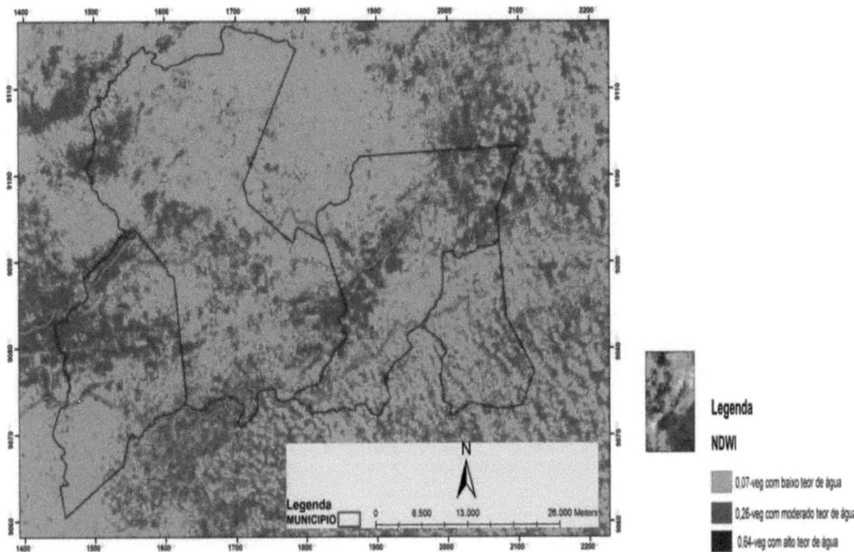

Figure 13 - NDWI for part of the middle section of the Ipojuca Basin in 1988.

The lowest values correspond to areas where soil is predominant, and the highest values correspond to areas where water is predominant. This is explained by the spectral response of the targets. Soils, in most cases, have higher reflectance values at 1.65pm than at 0.86pm. The opposite is true for vegetation and water, where the reflectance values at 1.65|jm are lower than at 0.86gm. Thus, the higher the reflectance value at 0.86|im, the higher the NDWI value (Gao, 1996).

To complement this NDWI result, the SAVI was also used, which is adjusted to the soil and introduces a factor into the NDVI to incorporate the effect of the presence of the soil. Using the SAVI, the fragmentation of the area's dense vegetation can be visualised (Figure 14).

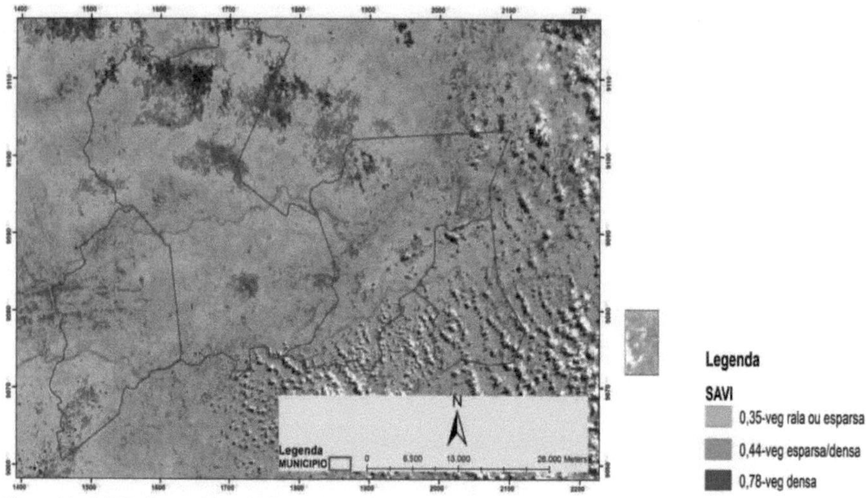

Figure 14 - SAVI of part of the middle section of the Ipojuca Basin in 1988.

It can be seen that the application of SAVI makes it much easier to differentiate between targets, i.e. to differentiate between exposed soil and sparse or sparse vegetation and also the difference between sparse/dense and dense vegetation. The next image analysed is from 28/02/1999, a period when the driest season is usually over and the rainy season begins in the region, which is why the areas of exposed soil can be seen better (Figure 15).

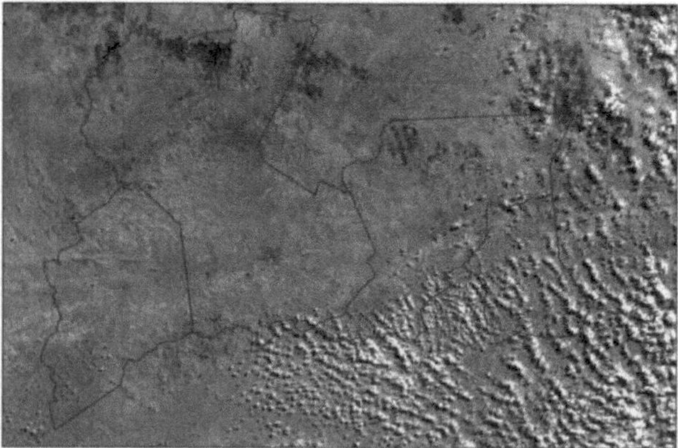

Figure 15 - Location of the municipalities in the middle section in 1999.

Figure 16 shows that in February 1999 (the start of the rainy season in this area) there was a greater presence of exposed soil and most of the vegetation had low or moderate water content. At this time of year, water stress in the vegetation has already reached its peak, which is why most of the image shows extensive areas with predominantly exposed soil.

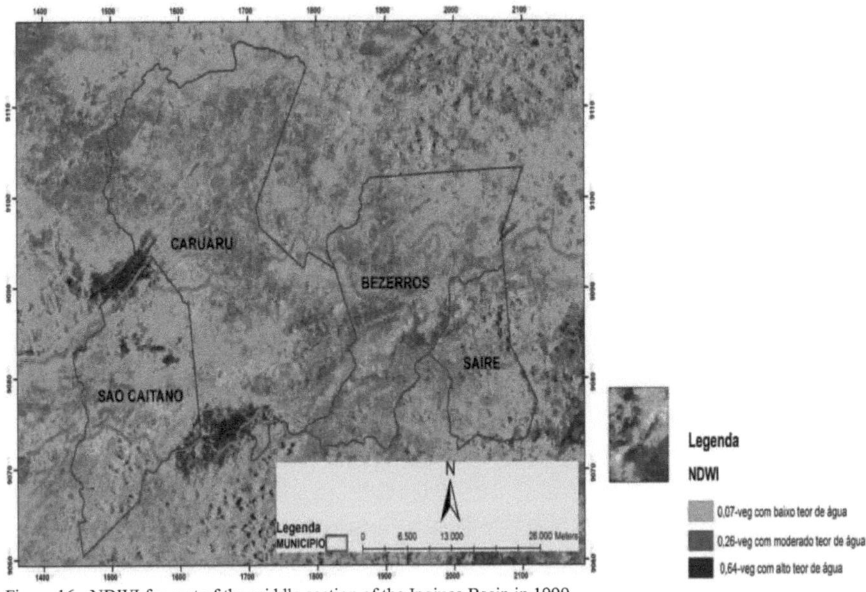

Figure 16 - NDWI for part of the middle section of the Ipojuca Basin in 1999.

Looking at Figure 13, it was possible to observe the behaviour of the NDWI in the area, and see that the areas where caatinga vegetation predominates, together with areas of exposed soil, are the ones with the most water stress. The highest NDWI values correspond precisely to the areas closest to the water bodies and with the greatest predominance of vegetation.

Similarly, the SAVI is very useful in this study, as it corroborates this result of low NDWI by also showing that at this time of year there would have been a decrease in areas with vegetation, with exposed soil predominating (in pink) and in the few areas where there was vegetation, it was sparse and/or sparse (Figure 17).

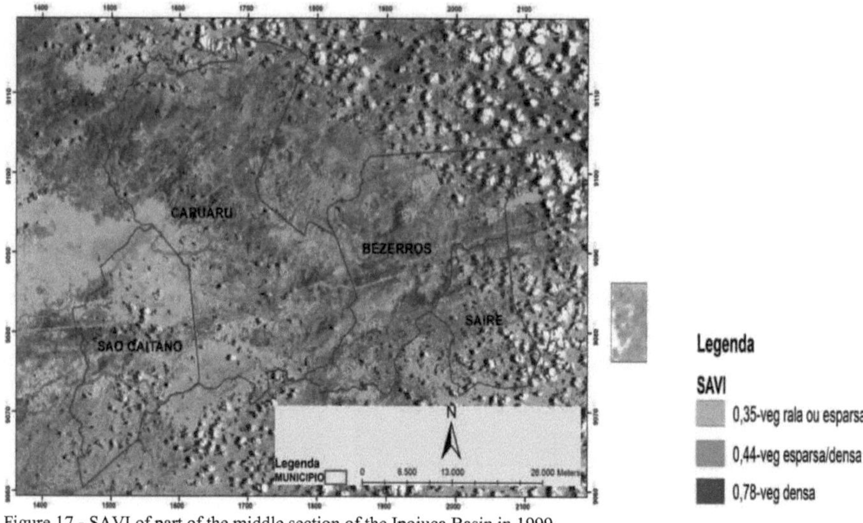

Figure 17 - SAVI of part of the middle section of the Ipojuca Basin in 1999.

It is important to point out that the application of SAVI to the 1999 image was very satisfactory as it showed a greater distinction between exposed soil and sparse vegetation.The next image to be analysed is from 19/07/2007, theoretically after the rainy season. The next image to be analysed is from 19/07/2007, theoretically after the rainy season. Looking only at the location of the area studied, it can be seen that, even without the SAVI vegetation index, compared to the rainy season in 1988, in 2007 the area has sparser vegetation, probably a reflection of the increase in land use and occupation in these areas (Figure 18).

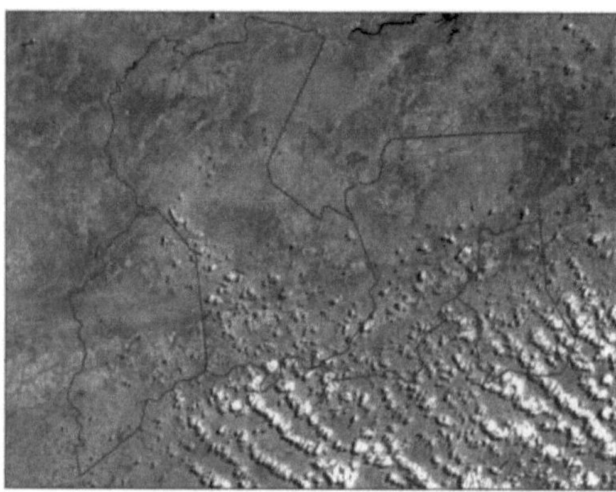

Figure 18 - Location of the municipalities in the middle section in 2007.

It can be seen that the area studied showed an increase in humidity, probably a reflection of the previous months of rain (Figure 19). As previously mentioned, the accumulation of water in the vegetation causes the NDWI values to increase, but there are still areas of exposed soil and sparse vegetation.

44

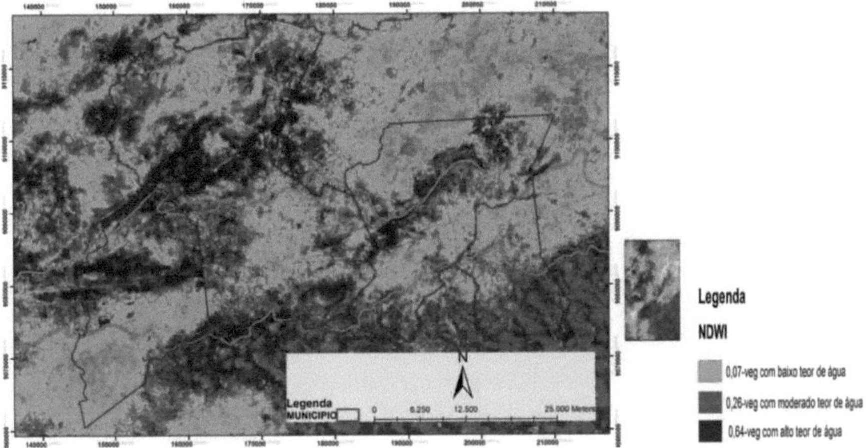

Figure 19- NDWI for part of the middle section of the Ipojuca Basin in 2007.

It is also important to note that this year's SAVI was able to better distinguish between sparse vegetation and denser vegetation, and that even with the increased humidity caused by the rainy season at this time of year, most of the vegetation in the area covered by the research was sparse (Figure 20).

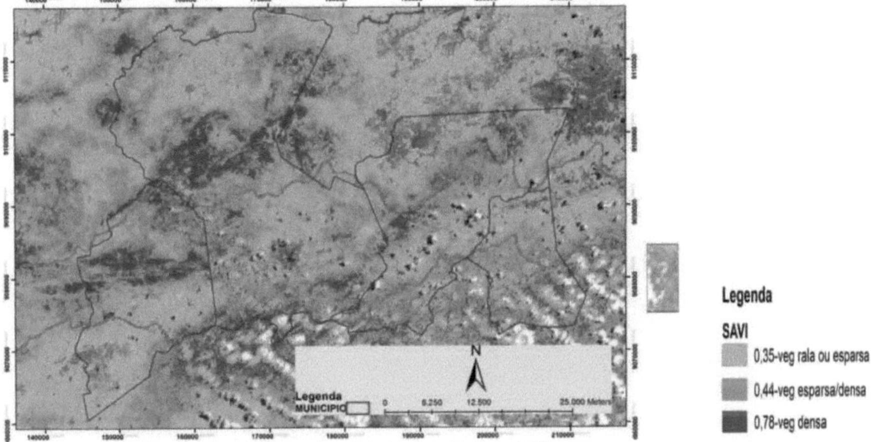

Figure 20 - SAVI of part of the middle section of the Ipojuca Basin in 2007.

In a more current image of the area, from 17/03/2011, despite the presence of some clouds, given the time of year, the study area can be clearly seen (Figure 21).

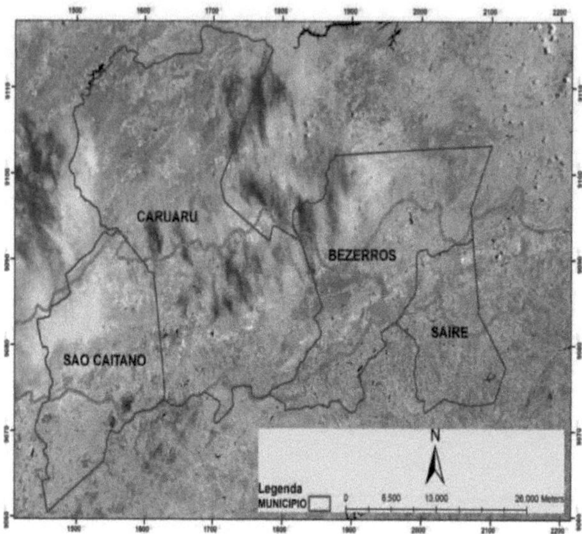

Figure 21 - Location of the municipalities in the middle section in 2011.

In the NDWI carried out in the area in 2011, the humidity appears to be well distributed, but there are still areas of exposed soil and the vegetation, in many areas, still has a low moisture content (Figure 22).

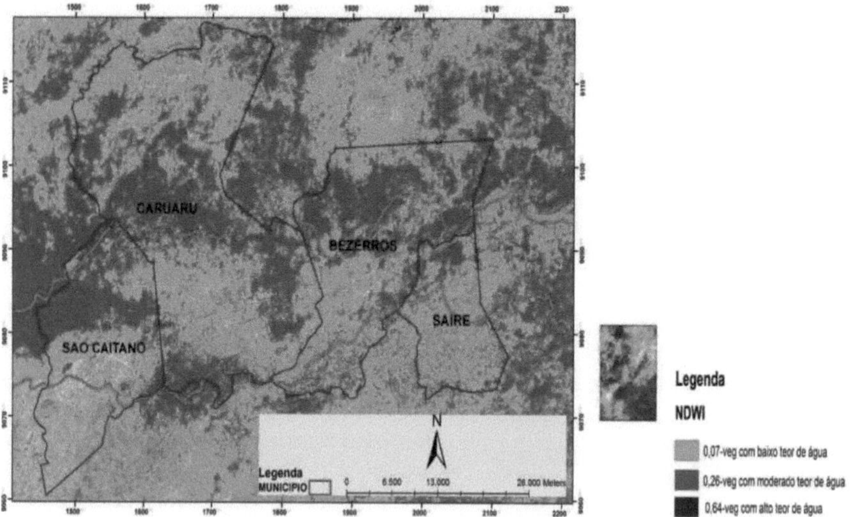

Figure 22 - NDWI for part of the middle section of the Ipojuca Basin in 2011.

Despite the humidity of the area, the vegetation is sparse (typical of areas where caatinga vegetation predominates), but it is more sparse this year (Figure 23).

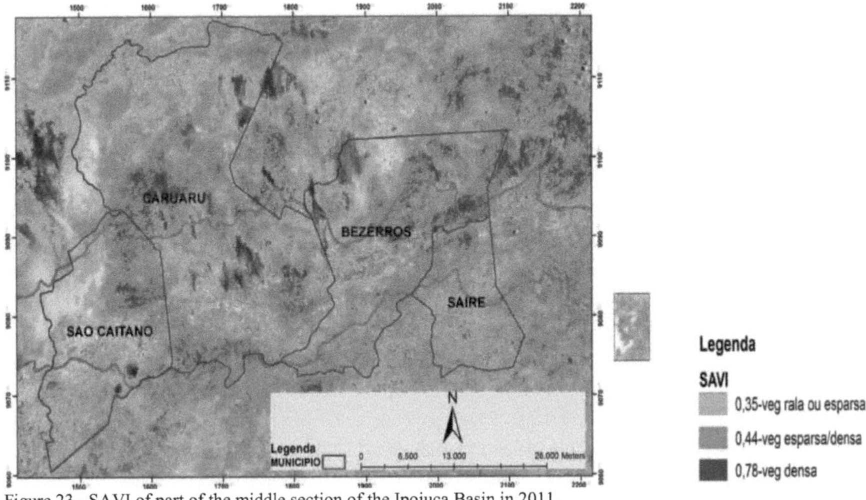

Figure 23 - SAVI of part of the middle section of the Ipojuca Basin in 2011.

It can be seen that in the SAVI index applied during the various years analysed, areas with exposed soil persist, especially in the lower right part of the municipality of São Caetano, even during periods of higher rainfall intensity. It is important to emphasise that these indices applied to images from different years serve to corroborate the results obtained with the MWSP methodology applied to the middle section of the Ipojuca Basin.

In this sense, using the moisture index (NDWI), it was noted that even during rainy periods the moisture content of the vegetation was moderate or low (in most of the area covered by the research) and that in these same periods, during different years of sampling, there were some areas of exposed soil which could potentially be susceptible to desertification processes.

These areas of exposed soil were constant from the first image (from 1988) to a more recent image (from 2011), and can be seen mainly in the municipality of São Caetano. In view of the above, there is a need for more in-depth studies of the existence of desertification centres in these areas, which are usually surrounded by sparse and/or sparse vegetation, and ways of avoiding or at least mitigating environmental impacts arising from the irrational use of the various resources of the region.

area and the possible environmental impacts of the works to transpose the São Francisco River.

According to Gutiérrez & Squeo (2004), when shrub cover is lost (by cutting, burning or grazing), the exposure of bare soil leads to the formation of a surface crust as a result of the direct impact of raindrops, which reduces water infiltration and increases surface runoff, reducing the chances of establishing vegetation cover. In this way, bare soil is highly susceptible to water and wind erosion, causing a net removal of nutrients from degraded areas.

In the areas affected by desertification in the semi-arid regions of north-eastern Brazil, the vegetation

47

is reduced in size and has greater spacing between its components than in other areas, generally coinciding with the presence of open hyperxerophilous caatinga. In this type of caatinga and soil, desertification can arise spontaneously, so there is a possibility that it existed in the Northeast before the colonisers arrived (VASCONCELOS SOBRINHO, 1982). It is precisely in order to better visualise the spacing of vegetation in the study area that the SAVI vegetation index was used.

According to Galindo et al. (2008) the fundamental characteristic of the desertification phenomenon in the northeastern semi-arid region is the presence of patches of exposed soil. These are generally areas of shallow soil with no water retention capacity and physical and chemical limitations, which increase the ecological vocation for desertification.

For Araújo Filho (2001), deserts begin to form with changes in the floristic composition of the plant community, causing the disappearance of the original vegetation, intensified by the gradual destruction of the floristic layer and the consequent reduction in the contribution of organic matter, which manifests itself in the soil with a loss of fertility and structure. These changes in the plant community can be caused by climate change or degradation due to anthropogenic activities, which would lead, respectively, to the climatic and ecological types of desertification mentioned by Conti (1998).

The natural dynamics of the Ipojuca Basin will be affected by the works to transpose the São Francisco River and by the flow of water from the implementation of this project. This will create new opportunities for land use and positive and/or negative impacts on nature in areas that did not receive water or received it at insufficient levels.

As such, the humidity and vegetation indices show that the middle section is a cause for concern. However, it is important to remember and emphasise that the situation in the upper section of the Basin must be much more worrying, due to the very climatic characteristics of the area located in the hinterland of the state of Pernambuco under the predominance of the semi-arid climate, thus inspiring future work.

Therefore, these studies provide support for both the methodological part and the actual results themselves, because in light of the above, it is clear that this Basin has shown a certain fragility with regard to its own flow dynamics and the maintenance of the natural characteristics of the caatinga vegetation, so climate change tends to accentuate the variation in flows as well as the effects on the population and economy of the surrounding areas. In addition to this, there are also the impacts of the transposition of the São Francisco River, which in the future will profoundly alter both the dynamics of flows and the forms of use and occupation of these areas, leading to new forms of land use and generating new spatialisations.

4.3 Discussion of the Possible Environmental Impacts of the Transposition of the São Francisco River.

According to the report prepared by the SBPC at the Workshop on the Transposition of the São Francisco River, the alteration of the flow regime and sediment flow resulting from the electricity sector projects implemented in the channel of the São Francisco River has generated major environmental impacts. Among these, the most significant have been the erosion of the estuary and surrounding areas, the formation of sandbanks in the lower reaches of the river, greater penetration of the salt wedge and the loss of fishing resources due to the lack of exchanges between the river and its marginal lagoons, which are breeding grounds for a large number of aquatic species.

The data collected at the workshop on the transposition of water from the São Francisco River indicated that most of the flow of the São Francisco River is compromised by energy generation activities and that there is a need to maintain a minimum flow in the lower reaches of the river to avoid further compromising environmental issues and productive activities such as fishing and navigation.

The technical, social and economic assessment of the two transposition axes proposed in the SBPC report, viewed more broadly, suggests that the Eastern Axis is more justifiable, given that it could solve the region's chronic problems. As for the Northern Axis, there are many more questions against its rationality in terms of real needs, social reach and economic and managerial viability.

Some of the main negative impacts cited in the RIMA were: changes to the ecosystems of the rivers in the receiving region, altering the population of aquatic plants and animals. The creation of aquatic environments distinct from the existing ones, the alteration of water volumes in the receiving rivers will promote a selection of species; risk of reducing the biodiversity of native aquatic biological communities in the receiving basins.

The selection between exotic and native species in the receiving regions could have an impact on the reduction of native species; the introduction of fish species that are harmful to humans in the region, such as piranhas and pirambebas, which feed on other fish and reproduce in still water; the reduction in the volumes of the reservoirs will lead to a reduction in fish biodiversity; some rivers do not have the capacity to receive the projected volume of water, flooding the parallel streams; and the deforestation of 430 hectares of land with native flora and the possible disappearance of the habitat of terrestrial animals that inhabit these regions. The most important species of flora are Caatinga Arbórea and Caatinga Arbustiva Densa.

Some of the other impacts mentioned in the report are: increased fluvial recharge of aquifers (positive impact), modification of the fluvial regime of the São Francisco River, loss of municipal revenues that are paid as compensation to the municipalities where the hydroelectric plants are concentrated, reduction in the generation of electricity on the São Francisco River, reduction in the diversity of terrestrial fauna, loss of land suitable for agriculture, instabilisation of the slopes around the water bodies, generation or increase of erosion and sediment transport during construction, initiation or acceleration of desertification processes during the operation of the system, alteration of the hydrosedimentological behaviour of the water bodies, risk of eutrophication of the new reservoirs, real estate speculation along the floodplains through which the canals will pass (ANNEX V).

One of the environmental impacts mentioned by the RIMA is the onset or acceleration of desertification processes during the operation of the system, which is particularly worrying for the upper and middle stretches of the Basin under study, since, according to the images worked on and through the application of the Australian methodology used, it was found, respectively, that there is high water stress and high vulnerability of the Basin to climate change in its middle section. This Basin may be susceptible in the coming years to possible climate variations and consequently to desertification processes caused both by the area's own climate dynamics and by anthropogenic action in land use and occupation and the deforestation of native vegetation.

4.4 Possible environmental impacts of the transposition of the São Francisco River in the middle section of the Ipojuca Basin.

The areas that are susceptible to desertification and fall within the scope of the United Nations Convention to Combat Desertification are those with arid, semi-arid and dry sub-humid climates. In this way, the area's climatic susceptibility can be assessed using the aridity index.

This study assessed the aridity index of some of the municipalities located in or near the middle section of the Ipojuca Basin. The aridity index was estimated using the ratio between average annual rainfall (P) and average annual potential evapotranspiration (ETP) over a 30-year period in certain municipalities that had their data studied, namely: São Caetano, Caruaru, Toritama and Belo Jardim, the latter two having replaced the municipalities of Bezerros and Sairé due to a lack of climatic and rainfall information from them.

It is important to note that the ETP estimates were based on long-period monthly averages obtained from the climatological normals provided by LAMEPE, and this procedure of evaluating long-period averages ends up ignoring the variations that occur on smaller time scales, such as annual and daily.

As an example of what was done in each municipality studied, in order to calculate the aridity index, the parameters relating to temperature, rainfall and potential evapotranspiration in the municipality of São Caetano were tabulated. The table shows the average monthly temperature for each month of the year (based on the monthly average established over a 30-year sample period), and also calculates the average evapotranspiration and precipitation over the same period of time. Equation 5 was used to calculate the Aridity Index, as shown in Table 21:

Table 21 - Aridity Index in the municipality of São Caetano.

SAO CAETANO			
T	ETP (mm/year)	P (mm/year)	IA
24,1	110,3	33	0,299072
24,1	102,0	52	0,509681
23,8	107,6	71	0,659909
23,4	97,6	70	0,717524
22,3	87,1	49	0,562886
21,1	71,7	61	0,850413
20,3	66,7	56	0,839497
20,5	69,1	24	0,347148

21,5	77,3	16	0,206987
22,9	96,3	1	0,01038
23,7	104,1	10	0,096097
24,1	113,8	23	0,20207
			0,441809

It should be noted that the water balance was calculated taking into account the altitude of the station and the coordinates of each municipality analysed. The aridity index data for each municipality analysed is shown in Table 22.

Table 22 - Aridity index of the municipalities in the study area.

MUNICIPALITIES	PRECIPITATION (mm)	TEMPERATURE (°C)	EVAPOTRANSPIRATION POTENTIAL (mm)	ARIDITY INDEX
BEAUTIFUL GARDEN	61	22,4	90,48	0,694
CARUARU	55,167	22,5	90,94	0,643
SAO CAETANO	38,833	22,6	91,97	0,442
TORITAMA	47,750	23,9	103,6	0,478

It can be seen from the results obtained by applying the aridity index that the municipalities of São Caetano and Toritama (which replaced the municipality of Bezerros in the data analysis and which is relatively close to it) are, according to the internationally accepted definition, included in a band characterised by the semi-arid domain with an aridity index varying between 0.21 and 0.50.

The municipality of Caruaru had an aridity index of 0.643 and is therefore in the dry sub-humid domain, which in the aridity index corresponds to the range between 0.41 and 0.65. The municipality of Belo Jardim (which replaced the municipality of Bezerros when the data was analysed) had an aridity index of 0.694 and, according to the proposed definition, falls within the humid sub-humid domain. In this way, it can be seen that these four municipalities are located in areas where, theoretically, there is a certain susceptibility to the desertification process and the data obtained by analysing the images relating to humidity and vegetation from the different years studied showed that in some areas the concern should be greater since, in addition to the susceptibility caused by the climate dynamics themselves, there are patches of exposed soil that have remained over the years and which are aggravated by the forms of land use and occupation, thus enabling the origin of nuclei where there is the start of a desertification process.

Another important point to analyse is the positive impact of the Transposition, which is the irrigation of abandoned areas and the creation of new agricultural frontiers. According to the land use capacity classes for annual and perennial crops, pastures, reforestation and wildlife, carried out by CONDEPE/FIDEM, Table 1 shows that most of the land in the upper and middle reaches of the Basin belongs to classes IV, V and VI, which are, respectively, land that can be cultivated only occasionally or to a limited extent and with serious conservation problems; land generally adapted for pasture and/or reforestation, with no need for special conservation practices, cultivable only in very special cases; and, land generally adapted for pasture and/or reforestation, with moderate conservation problems, cultivable only in special cases of a few soil-protecting permanent crops.

Given this reality, the transposition of the São Francisco River appears to be a viable and beneficial alternative for agriculture in the study area, with an increase in available water and a reduction in losses due to reservoirs, with a potential increase in agricultural production and consequently economic dynamism in the area, thus reducing the problems caused by drought, such as food shortages, low productivity in the countryside and rural unemployment. However, it is well known that along the floodplains through which the canals will pass and which have the greatest potential for agricultural activity, real estate speculation is expected to increase, so that the best land will remain in the hands of the few who can afford it.

However, it is important to be careful when observing the impacts that a large-scale project such as the Transposition can cause in certain areas, as it can be seen that in some areas of north-eastern Brazil there are already impacts from works carried out on this river, as an example cited by the São Francisco River Basin Committee, that after the regularisation of the river's flow, imposed by the operation of the hydroelectric plants, the advance of the sea on the north coast of Sergipe caused the destruction of two villages at the mouth of the São Francisco River, in 1998 and 2005, and this region is very sensitive to new interventions.

In view of this, studies such as this one, carried out in one of the receiving basins, such as the Ipojuca Basin, help to generate data and direct research, given the scarcity of data and bibliographic material, since the area covered by the impacts published in the RIMA comprises only a strip along the transposition channels, 5 km wide on each side. It is therefore necessary to carry out more complex studies of the dynamics of the receiving basins in order to assess the environmental impacts of the transposition project.

In this sense, the use of the moisture index (NDWI) showed that even during rainy periods, the moisture content of the vegetation was moderate or low (in most of the area covered by the research) and that in these same periods, during different years of sampling, there were some areas of exposed soil which could potentially be susceptible to desertification processes. These areas of exposed soil were constant from the first image (from 1988) to a more recent image (from 2011), and can be seen mainly in the municipality of São Caetano. In view of the above, there is a need for more in-depth studies of the existence of desertification centres in these areas, which are usually surrounded by sparse and/or sparse vegetation, and ways of avoiding or at least mitigating environmental impacts arising from the irrational use of the area's various resources and the possible environmental impacts of the São Francisco River transposition works.

Thus, using the humidity and vegetation indices, as well as the MWSP methodology, the middle section inspires caution. However, it is important to emphasise that the situation in the upper section of the Basin must be much more worrying, due to the climatic characteristics of the area, located in the hinterland of the state of Pernambuco and dominated by the semi-arid climate, thus inspiring future work.

Another important point to analyse is the positive impact of the Transposition, which is the irrigation of abandoned areas and the creation of new agricultural frontiers. According to the land use capacity classes for annual and perennial crops, pastures, reforestation and wildlife, carried out by CONDEPE/FIDEM, Table

1 shows that most of the land in the upper and middle reaches of the Basin belongs to classes IV, V and VI, which are, respectively, land that can be cultivated only occasionally or to a limited extent and with serious conservation problems; land generally adapted for pasture and/or reforestation, with no need for special conservation practices, cultivable only in very special cases; and, land generally adapted for pasture and/or reforestation, with moderate conservation problems, cultivable only in special cases of a few soil-protecting permanent crops.

Given this reality, the transposition of the São Francisco River appears to be a viable and beneficial alternative for agriculture in the study area, with an increase in available water and a reduction in losses due to reservoirs, with a potential increase in agricultural production and consequently economic dynamism in the area, thus reducing the problems caused by drought, such as food shortages, low productivity in the countryside and rural unemployment. However, it is well known that along the floodplains through which the canals will pass and which have the greatest potential for agricultural activity, real estate speculation is expected to increase, so that the best land will remain in the hands of the few who can afford it.

However, it is important to be careful when observing the impacts that a large-scale project such as the Transposition can cause in certain areas, as it can be seen that in some areas of north-eastern Brazil there are already impacts from works carried out on this river, as an example cited by the São Francisco River Basin Committee, that after the regularisation of the river's flow, imposed by the operation of the hydroelectric plants, the advance of the sea on the north coast of Sergipe caused the destruction of two villages at the mouth of the São Francisco River, in 1998 and 2005, and this region is very sensitive to new interventions.

Hence the concern about land use and occupation in the areas surrounding the Ipojuca Basin, because although the upper and middle stretches of the Basin are in great need of water, this Basin commonly has problems with pollution and flooding in its lower course.

CHAPTER 5

CONCLUSIONS

P As a preliminary study, the results were satisfactory in that they corroborated previous studies in the same area and identified new parameters and new methodologies for observing possible changes in a hydrological system.

C Considering the consumption of water for irrigation, the ratio of use of water resources obtained an average score.

The statistical evaluation of the flow data from the ipojuca River Basin showed that the area studied has a high degree of climatic variability.

In the statistical evaluation of the flow data, a result of high non-stationarity was obtained, thus denoting a significant tendency for the flow to increase or decrease over the series of years studied.

C By combining the data, a high level of vulnerability to climate change was found in the middle section of the Basin.

Based on the integration of data on vulnerability to climate change and the ratio of use of water resources, a high level of hydrological stress was found in the Ipojuca River Basin.

Given the results obtained in this study regarding the high hydrological stress in the Ipojuca Basin, and in addition to previous studies by other authors on the real conditions and future trends of changes in flow patterns in the Ipojuca Basin, there is growing concern about ways to avoid or at least mitigate possible impacts resulting from these changes.

The statistical analysis of the NDWI images showed a low soil moisture index in parallel with high water stress in the vegetation in the years studied, even in the rainiest periods.

The application of the SAVI vegetation index to the images to assess the level of biomass showed the growth, over the years, of areas with a lower density of vegetation.

The results obtained of high water stress in the vegetation suggest that the transposition of the waters of the São Francisco River is the most appropriate alternative for the reality of the study area.

T However, in light of the Environmental Impact Report (RIMA) carried out by the Ministry of National Integration, it can be seen that the benefits brought to the population living in the areas covered by the project, and to the environment, fall short of the negative impacts caused by the project.

By calculating the aridity index, it was observed that the areas studied in the middle section of the Ipojuca Basin are located in areas that, according to the United Nations Convention to Combat Desertification, are susceptible to desertification processes.

It is advisable to continue this work by expanding the scope of observation of the study area and the number of stations studied in order to have a more general view of the territory, taking into account the upper, middle and lower reaches of the Ipojuca Basin and not just parts of it, given that this is one of the basins that will receive water from the transposition of the São Francisco River.

REFERENCES

ALLAN, J. D. 1995. **Stream ecology:** structure and function of running waters. Chapman and Hall, London, 388p.

ALLEN, R.G.; TASUMI, M.; TREZZA, R. 2002. **SEBAL (Surface Energy Balance Algorithms for Land).** Advance Training and Users Manual - Idaho Implement. 97p.

ALVES, M. A.; HENRIQUES, A. G. 1994. **Ecological flow as a minimisation measure:** methods for determining it. In: Actas do 6° SILUSB/1°SILUSBA, Simpósio de hidráulica e recursos hídricos dos países de Língua Portuguesa, Lisbon, APRH/ABRH, pp. 177-190.

ANA, 2004. **ANA Technical Note No. 600/2004/SOC**. Brasilia, DF. Optical medium. 14p.

ARAÚJO FILHO, J. A. Our daily desertification. Rumos & Debates Collection. Available at: http://www.embrapa.br.:8080/aplic/rumos.nsf/f7c8b9aeabcc8583. Accessed on: 01 August 2011.

BASTOS, A. C. S.; ALMEIDA, J. R. **Brazilian environmental licensing in the context of environmental impact assessment.** In: Cunha; Guerra (org). Avaliação e perícia ambiental. 7. ed. Rio de Janeiro: Bertrand Brasil, 2006.

BRANDÃO, S.L.; LIMA, S.C. **Diagnóstico ambiental das áreas de preservação permanente (APP), margem esquerda do Rio Uberabinha, em Urbelândia (MG).** In: caminhos de geografia - online magazine. October 2002.

Cardozo, F.S.; Pereira, G.; Silva, G.B.S.S.; Silva, F.B.; Shimabukuro, Y.E.; Moraes, E.C. 2009. Discrimination of flooded areas in the Pantanal of Mato Grosso do Sul using orbital images. Anais 2° Simpósio de Geotecnologias no Pantanal, Corumbá, 7-11 November, Embrapa Informática Agropecuária/INPE, p.99-106.

CASTRO, P. R. C. Citrus behaviour under water deficit. **Orange**. v. 2, n.15, p.139- 154, 1994.

CONAMA. **CONAMA Resolutions**, 1986 to 1991. Brasília: IBAMA, 1992.

CONTI, J.B. 1998. Climate and the environment. São Paulo: Atual. 88p.

COSTA, M. V.; CHAVES, P. S. V.; OLIVEIRA, F. C. Uso das Técnicas de Avaliação de Impacto Ambiental em Estudos Realizados no Ceará. **XXVIII Brazilian Congress of Communication Sciences.** Intercom

- Brazilian Society for Interdisciplinary Communication Studies. Uerj, 2005.

CHANDER, G.; MARKHAM, B. 2003. Revised Landsat-5 TM Radiometric Calibration Procedures and Postcalibration Dynamic Ranges. **IEEE TRANSACTIONS ON GEOSCIENCE AND REMOTE SENSING.** v. 41. n. 11.

CHAVES, H.M.L; ROSA, J.W.C.; SANTOS, V.M. 1997. Evaluation of the trapping efficiency of Gallery Forests through sedimentation modelling. In: Imana-Encinas, J. & Christoph Kleinn (eds.). **Proceedings:** the international symposium on assessment and monitoring of forests in tropical dry regions with special reference to gallery forests. Brasília: University of Brasília. 378p. : il.

CLARKE, R. T. 2003. Frequency analysis of hydrological events and the use of long memory models. In: Tucci, C.E.M. and Braga, B. (eds). **Climate and water resources in Brazil.** ABRH, Porto Alegre, RS, pp. 243-264.

CONDEPE. 2005. **Ipojuca River Basin:** Pernambuco Hydrographic Basins Series. Pernambuco State Government, Planning Secretariat, Pernambuco State Planning and Research Agency. 64 pages.

EMBRAPA, 2010. **Diagnosis and alternatives for the environmental recovery of the Basin Guandu River Hydrographic Basin** (BHRG) - RJ, 2010

FENSHOLT, R.; SANDHOLT, I. Derivation of a shortwave infrared water stress index from MODIS near- and shortwave infrared data in a semiarid environment. **Remote Sensing of Environment,** v. 87, p. 111-121, 2003.

FREITAS, C. M.; PORTO, M. F. S. 2006. **Health, environment and sustainability.** Rio de Janeiro: Ed. Fiocruz.

FREITAS, M. A V. & SANTOS. A H. M. (2000). **Perspectives on Water Resources Management and Information.** In: **The State of Water in Brazil,** ANEEL - Brasília - DF, Part 1.

GALINDO, I.C.L.; RIBEIRO, M.R.; SANTOS, M.F.A.V.; LIMA, J.F.W.F. & FERREIRA, R.F.A.L. Soil-vegetation relations in areas under desertification in the municipality of Jataúba, PE. R. Bras. Ci. Solo, 32:1283-1296, 2008.

GALVÃO, D. M. O. 2008. **Subsidies for determining environmental flows in unregulated watercourses:** the case of Ribeirão Piripau (DF/GO). Master's dissertation, University of Brasilia, Department of Forestry Engineering, Publication PPGEFL. DM - 096/08, Brasília, DF.

GAO, B.C. 1996. NDWI - A normalised difference water index for remote sensing of vegetation liquid water from space. **Remote Sensing of Environment,** v. 58, p. 257-266.

GONÇALVES, M. V. C. 2003. **Methodology for determining minimum guaranteed flows in watercourses.** Master's dissertation, University of Brasília, Department of Civil and Environmental Engineering, Publication MTARH-DM - 061/03, Brasília, DF, 129p.

GUTIÉRREZ, J.R. & SQUEO, F.A. Importance of shrubs in semi-arid ecosystems in Chile. Ecosistemas, 2 0 0 4 / 1 Disponively: <URL:http//www.aeet.org/ecosistemas/041/investigacion2.htm>

HUETE, A. R. Adjusting vegetation indices for soil influences. **International Agrophysics,** v.4, n.4, p.367-376, 1988.

IPCC (Intergovernmental Panel on Climate Change) 2001. **Impacts, Adaptation and Vulnerability.** J.

M. McCarthy et al., editors. Cambridge University Press, Cambridge, UK, 1032 pp.

IPCC (Intergovernmental Panel on Climate Change) 2007. **New Climate Scenarios**. IPCC/UN Report.

MARKHAM, B.L.; BARKER, L.L. 1987. Thematic mapper bandpass solar exoatmospherical irradiances. **International Journal of Remote Sensing**, v.8, n.3, p.517-523.

NEW SOUTH WALES (2006). **Macro Water Sharing Plans:** the approach for unregulated rivers. Australia: Department of Natural Resources - NSW. 56p.

NOVO, E. M. **Remote sensing: principles and applications**. São Paulo. Edgard Blucher, 1989. 309p.

OLIVEIRA, T. H.; SILVA, J. S.; MACHADO C. C. C.; GALVÍNCIO, J. D.; PIMENTEL, R. M. M.; SILVA, B. B. 2010. Moisture index (NDWI) and spatio-temporal analysis of surface albedo in the Moxotó river basin. **Brazilian Journal of Physical Geography.** Homepage: www.ufpe.br/rbgfe.

PEREIRA, A. R.; VILLA NOVA, N. A.; SEDIYAMA, G. C. **Evapo(transpi)rtion**. Piracicaba: FEALQ, 1997. 183p.

PERNAMBUCO. 2004. Diagnosis of water resources in the GL-2 basin, consolidation of existing studies, preparation of the Water Resources Utilisation Plan for the Metropolitan Region of Recife, Zona da Mata and Agreste **Pernambucano** and integrated water resources management model: **Water Resources Utilisation Plan for the Metropolitan Region of Recife, Zona da Mata and Agreste Pernambucano** - Volume III Water Demand Studies for the Region Studied. Secretariat for Science, Technology and the Environment, PROÁGUA Semiárido, TECHNE Engenheiros Consultores, Recife-PE, 266p.

PIMENTEL, G.; PIRES, S. H. Metodologias de avaliação de impacto ambiental: Aplicações e seus limites. Rio de Janeiro, **Revista de Administração Pública,** 26 (1), p.56-68, 1992.

PNMA, 2003. **Updating and complementing water uses in the Ipojuca River Basin.** Project: Water Quality Monitoring as a Tool for Environmental Control and Water Resource Management in the State of Pernambuco, Recife, PE, 13p.

REIS, J.A.T; GUIMARÃES M.A.; BARRETO NETO, A.A. BRINGHENTI, J. 2008. Regional indicators applicable to assessing the flow regime of watercourses in the Itapoana River Basin. **Geociências** (São *Paulo)* [online]. 2008, vol. 27, n° 4, pp. 509516. ISSN 0101-9082.

ROSA, R. BRITO, J. L. S. **Introduction to geoprocessing:** geographic information systems. Uberlândia: Editora da Universidade Federal de Uberlândia, 1996. 104p.

ROSS, J. L. S. O Registro Cartográfico dos Fatos Geomórficos e a Questão da Taxonomia do Relvo. In: **Revista do Departamento de Geografia** - FFLCH-USP, n° 6, São Paulo, 1992.

ROSS, J.L.S. - **Geomorphology:** environment and planning. 6th edition, SP, Contexto, 2001.

SALAS, J.D. 1993. **Analysis and modelling of hydrologic time series**. In: Maidment, D.R. (ed.) Handbook of hydrology. MacGraw-Hill, United States of America, pp. 19.1-20.1.

SÁNCHEZ, L. E. **Environmental impact assessment:** concepts and methods. São Paulo : Oficina de Textos, 2006.495 p.

SILVA, A. M. da. **Environmental impact study**: ecological planning. João Pessoa: SUDEMA, 1989.

SILVA, E. R. A. C.; VIDAL, C. V. S.; SILVA, C. A. V.; GALVÍNCIO, J. D. 2009. Preparation of a land use and occupation map of the Suape harbour area using images from the HRC and CCD sensors of the CBERS 2B satellite. **III Colloquium on Technology and Geosciences.** UFPE, 2009.

SILVA, I. F.; NÓBREGA, R. S.; GALVÍNCIO, J. D. 2009. Impact of Climate Change on the Hydrological Responses of the Ipojuca River (PE) - Part 2: Temperature Increase Scenarios. **Revista Brasileira de Geografia Física**. 2009, vol. 2, n°2, pp. 19-30.

SOUZA FILHO, F. A. 2003. Climate variability and change in semi-arid Brazil. In: TUCCI, C.E.M. and Braga, B. (eds.). **Climate and water resources in Brazil.** ABRH, Porto Alegre, RS, pp. 77-116.

TUCCI, C.E.M. 1998. **Hydrological Models.** Editora Universitária UFRGS, Porto Alegre.

TUCCI, C.E.M. 2002. **Regionalisation of flows.** University Press. UFRGS. 1st edition. Porto Alegre.

TUCCI, C.E.M.; BRAGA, B. 2003. Climate and water resources. In: Tucci, C.E.M. and Braga, B. (eds). **Climate and water resources in Brazil**. ABRH, Porto Alegre, RS, pp. 1-22.

TUCCI, C. E. M.; CLARKE, R. T., 2003. Hydrological regionalisation. In: Paiva, J. B. D. and Cauduro, E. M. (eds). **Hydrology applied to the management of small river basins.** ABRH, Porto Alegre, RS, pp. 169-222.

TUCCI, C.E.M.; MENDES, C.A., 2006. **Integrated environmental assessment of river basins**. MMA/SQA, Brasília, DF, 300p.

UNESCO. **Water:** a shared responsibility. 2006. The United Nations World Water Development Report 2, UNESCO, Oxford-UK, 584p.

VERDUM, R.; MEDEIROS, R. M. V. (org.). **RIMA: environmental impact report. Legislation, preparation and results.** Porto Alegre: Editora UFRGS, 2002. 4ª ed. expanded version. 2002.

VASCONCELOS SOBRINHO, J. Desertification processes occurring in the Northeast of Brazil: Their genesis and containment. Recife, SEMA/SUDENE, 1982. 101p.

ANNEXES

ANNEX I - Mesoregions, Microregions, Development Regions, Municipalities and Districts Drained by the Ipojuca Basin.

Mesoregions	Microregions	Development Regions (DR)	Municipalities and Districts	Municipalities		
				Total (km)^2	Belonging to the Basin	
					km^2	%
Hinterland Pernambuco	Moxotó Hinterland	RD Sertão do Moxotó	Arcoverde	380,60	104,09	27,35
Agreste	Ipojuca Valley	RD Agreste Central	Alagoinha Perpétuo Socorro*	180,10	54,61	30,32
			Belo Jardim(*) Água Fria(*)	653,60	230,92	35,33
			Bezerros(*) Boas Novas(*)	545,70	226,96	41,59
			Cachoeirinha	183,20	1,81	0,99
			Caruaru(*)	932,00	387,62	41,59
			Gravatá(*) Mandacaru(*)	491,50	169,03	34,39

			Pesqueira Mutuca(*) Papagaio(*) Salobro(*)	1.036,00	606,79	58,57
			Poção(*) Poção Sugar Loaf(*)	212,10	189,62	89,40
			Riacho das Almas	313,90	8,19	2,61
			Sanharó(*) Mulungu(*)	247,50	235,45	95,13
			São Bento do Una	715,90	70,15	9,80
			São Caetano(*) Maniçoba(*)	373,90	262,37	70,17
			Tacaimbó(*) Riacho Fechado(*)	210,90	131,81	62,50
			Venturosa	326,10	2,22	0,68
	Ipanema Valley	RD Agreste Meridional	Altinho Ituguaçu(*)	452,60	6,70	1,48
			Sairé	198,70	75,88	38,19
	Victory of St Antony	RD Mata Sul	Chã Grande(*)	83,70	68,52	81,86
			Pigeons Our Lady of Carmo(*)	236,10	66,51	28,17
			Victory of St Antony	345,70	39,79	11,51
Mata Pernambuco	Pernambuco's Southern Forest		Amaraji	238,80	60,89	25,50
			Ladder(*)	350,30	203,73	58,16
			Spring(*)	96,50	79,09	81,96
Metropolitan Recife	Suape	RD Metropolitan	Ipojuca(*)	514,80	150,84	29,30
Total						3.433,58

Source: **Ipojuca River Basin Master Plan / IBGE / CONDEPE/FIDEM Agency.**
(*) Municipality/District with headquarters in the basin.

ANNEX II - Resident Population and Area of the Municipalities Drained by the Basin, 2004.

Municipalities	Population (inhabitants)			Municipal area(**) (km)^2	Population density (inhabitants/km)^2
	Total	Urban	Countryside		
Alagoinha	13.288	7.627	5.661	200,42	66,30
Altinho	21.611	11.821	9.790	454,49	47,55
Amaraji	21.982	17.014	4.968	234,78	93,63
Arcoverde	64.588	58.323	6.265	353,38	182,77
Beautiful Garden(*)	72.823	53.816	19.007	647,70	112,43
Calves(*)	60.058	49.248	10.810	492,56	121,93
Cachoeirinha	17.653	12.887	4.766	179,27	98,47
Caruaru(*)	274.124	235.747	38.377	920,61	197,76
Chã Large(*)	19.899	14.566	5.333	70,19	283,50
Ladder(*)	58.111	49.801	8.310	347,20	167,37

59

Gravatá(*)	70.243	60.549	9.694	513,37	136,83
Ipojuca(*)	66.390	42.357	24.033	527,32	125,90
Fishing	57.772	41.769	16.003	1.000,22	57,76
Potion(*)	11.996	7.174	4.822	199,74	60,06
Pigeons	24.429	16.074	8.355	207,66	117,64
Spring(*)	11.797	7.491	4.306	109,94	107,30
Riacho das Almas	18.245	6.878	11.367	313,99	58,11
Sairé	14.950	7.625	7.325	195,46	76,49
Sanharó(*)	16.318	7.931	8.387	256,18	63,70
São Bento do Una	46.963	25.360	21.603	726,96	64,60
São Caetano(*)	35.390	25.764	9.626	382,48	92,53
Tacaimbó(*)	13.572	6.772	6.800	227,59	59,63
Venturosa	14.176	8.761	5.415	338,12	41,93
Vitória de St°. Antão	123.130	106.631	16.499	371,80	331,17
Total	**1.149.508**	**881.986**	**267.522**	-	-

Source: IBGE/Agência CONDEPE/FIDEM.
(*) Municipality with headquarters in the basin.
(**) Area of municipalities according to IBGE Resolution No. 05 of 10 October 2002.

ANNEX III - Donor and Recipient Basins:

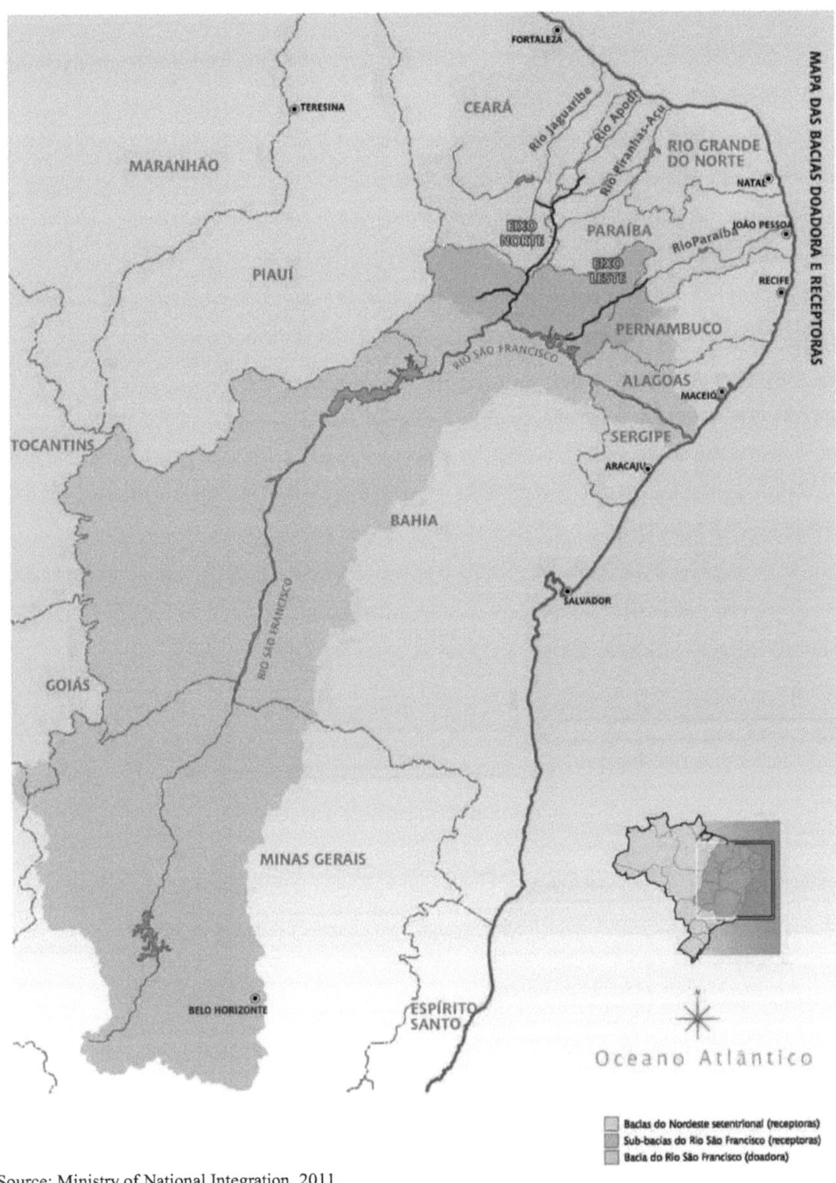

Source: Ministry of National Integration, 2011.

ANNEX IV - Part of the Environmental Impact Report (RIMA) for the São Francisco River Basin:

- Positive impacts:

1. Increased water availability and reduced loss from reservoirs.

2. Generation of 5,000 jobs during construction (four years), especially in the towns where the construction sites will be located. However, at the end of construction, there will be no significant impact in terms of job creation.

3. Increased income and trade in the affected regions. During the work, there will be a big increase in trade and income in the cities that will host the construction sites. In the long term, the increase in employment and income will come from the irrigated agriculture and industry that will result from the transposition.

4. Supplying up to 12.4 million people in cities, through urban supply systems that have already been implemented, are being implemented or are being planned by local authorities.

5. Rural supply with good quality water. The project envisages the construction of public fountains in 400 urban locations in the sertão region that do not have an adequate water supply system.

6. Reduction of problems caused by drought, such as food shortages, low productivity in the countryside and rural unemployment. 340,000 people would benefit, especially in the Piranhas-Açu basin (39%) and the Jaguaribe basin (29%).

7. Irrigation of abandoned areas and creation of new agricultural frontiers. According to the studies carried out, approximately 161,500 hectares could be made possible by 2025, 24,400 hectares of which for diffuse irrigation along the canals and 137,100 hectares for planned irrigation.

8. The water quality of the rivers and reservoirs in the receiving regions will benefit from the waters of the São Francisco.

9. The supply of water will help to keep around 400,000 people in the countryside.

10. Reduction in illnesses and deaths caused by consumption of contaminated water or lack of water. It is estimated that the number of hospitalisations caused by water-related diseases will fall by around 14,000 in 2025, compared to a forecast of 53,000 in the absence of the project.

11. Reduced pressure on the health infrastructure due to fewer cases of diseases caused by unsafe water.

- Negative impacts:

1. Loss of employment for the population in the expropriated areas and for the workers at the end of the works.

2. Modification of the ecosystems of the rivers in the receiving region, altering the population of aquatic plants and animals. The creation of aquatic environments that are different from the existing ones, the alteration of water volumes in the receiving rivers will promote a selection of species. Fish and other aquatic organisms are important in reconstructing the biogeographical history of river basins. Altering ecosystems can have an impact on our knowledge of the region's history.

3. Risk of reducing the biodiversity of native aquatic biological communities in the receiving basins. Selection between exotic and native species in the receiving regions could have an impact on the reduction of native species.

4. Introduction of social tensions and risks during the construction phase. At the start of the works, job and income losses are expected in rural areas due to expropriations, the removal of the population from the regions where the canals will pass, immigration to the cities in search of work on the construction sites. At the end of the works, the dismissal of workers could be a source of conflict.

5. The expropriation of land and the exodus from the affected regions will alter the way of life and community ties of kinship and friendship, which are very important for coping with the precarious living conditions of many communities.

6. Movement of workers through indigenous lands of two ethnic groups: Truká and Pipipã, generating undesirable interference.

7. Pressure on urban infrastructure in the cities that will receive the workers, increasing the demand for housing and health services. Rising reservoir levels can cause water-related diseases such as dengue fever and schistosomiasis. Contact with construction workers can increase cases of sexually transmitted diseases.

8. The project region has many archaeological sites, putting them at risk of losing this heritage due to excavations in the areas to be flooded by the reservoirs and in the course of the rivers whose volume will be increased.

9. Deforestation of 430 hectares of land with native flora and the possible disappearance of the habitat of the terrestrial animals that inhabit these regions. The most important species of flora are Caatinga Arbórea and Caatinga Arbustiva Densa.

10. Introduction of fish species harmful to humans in the region, such as piranhas and pirambebas, which feed on other fish and breed in still water.

11. The decrease in weir volumes will lead to a reduction in fish biodiversity.

12. Some rivers do not have the capacity to receive the projected volume of water, flooding the parallel streams.

ANNEX V - Other impacts mentioned in the Environmental Impact Report (RIMA):

Of the other 21 impacts included in the report, only the first one listed below is considered positive. The others are classified as negative, according to page 75 of the RIMA. They are:

1. Increased river recharge of aquifers.

2. Changes in the fluvial regime of the São Francisco River.

3. Reduction of electricity generation on the São Francisco River.

4. Loss of municipal revenues that are paid as compensation to the municipalities where the hydroelectric plants are located.

5. Fish and other aquatic organisms are important in reconstructing the biogeographical history of river basins. The alteration of ecosystems can have an impact on our knowledge of the region's history.

6. An increase in hunting activities and a decrease in the population of cynergetic species due to deforestation during the construction phase. The animals threatened by these activities are amphibians, reptiles, mammals and birds. Some of these animals are vulnerable or threatened with regional extinction, such as the ball armadillo, the jaguar, the capuchin monkey, the armadillo, the bush pig and the soft-tailed armadillo.

7. Decrease in the diversity of terrestrial fauna.

8. Loss of land suitable for agriculture.

9. Instabilisation of slopes around bodies of water.

10. Generation or increase of erosion and sediment load during construction.

11. Start or acceleration of desertification processes during system operation.

12. Alteration of the hydrosedimentological behaviour of water bodies.

13. Risk of eutrophication of the new reservoirs.

14. Risk of accidents to the public during the work due to the transit of machinery and equipment.

15. Increased dust emissions during the construction and operation of the system.

16. Conflicts in the mining areas through which the water will pass.

17. Property speculation along the floodplains through which the canals will pass.

18. Risk of accidents with venomous animals, especially snakes.

19. Increase and/or appearance of diseases: The rise in the level of reservoirs and water in rivers can cause water-related diseases such as dengue fever and schistosomiasis. Contact with construction workers can increase cases of sexually transmitted diseases.

20. Risk of vector proliferation: canals, reservoirs and weirs are favourable environments for the host of schistosomiasis and vectors of dengue, malaria and yellow fever.

21. The spread of the above diseases can put pressure on health services in the affected region.

Printed by Books on Demand GmbH, Norderstedt / Germany